OXFORD WORLD'S CLASSICS

ON WAR

CARL VON CLAUSEWITZ was born in Burg near Magdeburg in Prussia in 1780. The son of a low-ranking Prussian officer, he was educated as a cadet in the army, and was on active military service in his teens, fighting against France in the French Revolutionary Wars. He studied military history under Gerhard von Scharnhorst, whom he far surpassed intellectually, developing his knowledge of the tactics employed in the great European conflicts of the eighteenth century. In 1803 he met Marie, Countess von Brühl, who became his wife in 1810.

Clausewitz took part in the battles of Jena and Auerstedt in 1806, when Napoleon defeated Prussia, and was briefly imprisoned by the French. On his release he joined Scharnhorst's commission into the Prussian military, instituting wide-ranging reforms. He witnessed the battles of Smolensk and Borodino during Napoleon's 1812–13 campaign, fighting briefly on the Russian side, which provided valuable insights for his future work. Clausewitz became the administrative director of the Military Academy in Berlin in 1815, and it was there, between the years 1815 and 1830, that he wrote *On War*. In 1830 he was sent to prevent a Polish insurrection; he died of cholera on 16 November 1831. *On War* was published by his widow in 1832.

MICHAEL HOWARD is Emeritus Professor of Modern History at both Oxford and Yale universities. PETER PARET is Professor Emeritus at the Institute of Advanced Studies, Princeton University.

BEATRICE HEUSER is a Professor at the University of the German Federal Armed Forces, Munich. Her previous positions include Professor of International and Strategic Studies in the Department of War Studies at King's College London, and Director of Research at the Military History Research Institute in Potsdam. Her publications include *Reading Clausewitz* (2002).

OXFORD WORLD'S CLASSICS

*For over 100 years Oxford World's Classics have brought
readers closer to the world's great literature. Now with over 700
titles—from the 4,000-year-old myths of Mesopotamia to the
twentieth century's greatest novels—the series makes available
lesser-known as well as celebrated writing.*

*The pocket-sized hardbacks of the early years contained
introductions by Virginia Woolf, T. S. Eliot, Graham Greene,
and other literary figures which enriched the experience of reading.
Today the series is recognized for its fine scholarship and
reliability in texts that span world literature, drama and poetry,
religion, philosophy and politics. Each edition includes perceptive
commentary and essential background information to meet the
changing needs of readers.*

OXFORD WORLD'S CLASSICS

═══

CARL VON CLAUSEWITZ

On War

═══

Translated by
MICHAEL HOWARD *and* PETER PARET

Abridged with an Introduction and Notes by
BEATRICE HEUSER

OXFORD
UNIVERSITY PRESS

OXFORD

UNIVERSITY PRESS

Great Clarendon Street, Oxford OX2 6DP

Oxford University Press is a department of the University of Oxford.
It furthers the University's objective of excellence in research, scholarship,
and education by publishing worldwide in

Oxford New York

Auckland Cape Town Dar es Salaam Hong Kong Karachi
Kuala Lumpur Madrid Melbourne Mexico City Nairobi
New Delhi Shanghai Taipei Toronto

With offices in

Argentina Austria Brazil Chile Czech Republic France Greece
Guatemala Hungary Italy Japan Poland Portugal Singapore
South Korea Switzerland Thailand Turkey Ukraine Vietnam

Oxford is a registered trademark of Oxford University Press
in the UK and in certain other countries

Published in the United States
by Oxford University Press Inc., New York

British Library Cataloguing in Publication Data

Data available

Library of Congress Cataloging in Publication Data

Clausewitz, Carl von, 1780–1831.
[Vom Kriege. English]
On war / Carl von Clausewitz ; translated by Michael Howard and Peter Paret ;
abridged with an Introduction and Notes by Beatrice Heuser.
p. cm. — (Oxford world's classics)
Translation of *Vom Kriege*
Translation originally published: Princeton University Press, c1976.
Includes bibliographical references and index.
ISBN–13: 978–0–19–280716–8 (alk paper)
ISBN–10: 0–19–280716–1 (alk paper)
1. Military art and science. 2. War. I. Howard, Michael Eliot, 1922– II. Paret, Peter.
III. Heuser, Beatrice, 1961– IV. Title.
U102.C65 2006
355.02—dc22
2006019812

ISBN 978–0–19–954002–0

5

Typeset in Ehrhardt
by RefineCatch Limited, Bungay, Suffolk

Printed in Great Britain on acid-free paper by Clays Ltd, St Ives plc

CONTENTS

INTRODUCTION

PERHAPS the most intriguing question for a reader first confronted with the work of Carl von Clausewitz is what made *On War* so special. Why is a book of military philosophy written more than 175 years ago still so influential today? The answers are many and various, and lie partly in the difference between *On War* and its predecessors on the subject, and in the unique intellectual skills of Clausewitz himself. Gerhard von Scharnhorst's *Handbook for Officers for Use in the Field*, first published in 1793, typified the approach of previous writing on war. Scharnhorst (1755–1813), the director of the Academy for Young Officers which Clausewitz attended, was a lifelong influence on the younger man. Clausewitz called him the 'father of my mind' and held him in great esteem, yet the differences between their two books could not have been greater. Scharnhorst's *Handbook* stands in a clear line of manuals on the art of war in the tradition of the Roman military writer Flavius Vegetius' *Epitoma rei militaris* of the late fourth century, which in turn drew on many classical works subsequently forgotten in the Middle Ages. The works of Vegetius and of many subsequent authors such as Don Bernardino de Mendoza, Diego de Salazar, Marshal Raimondo Montecuccoli, the chevalier de la Valière, the marquis de Feuquières, the chevalier de Folard, Marshal Maurice de Saxe, the Puységurs (father and son), King Frederick the Great of Prussia, General Henry Lloyd, Archduke Charles, and even the politically minded Machiavelli in his *Art of War* were in many ways mere cookery books. They were all divided into many chapters, sections, and subsections, giving precise and unequivocal rules to follow on anything from the criteria for the appointment of a good captain to the amount of food required for each soldier, from how to conduct marches at night to how to dig temporary or more permanent trenches around campsites, from how to invest fortified places to how to attack in battle with light cavalry. As Scharnhorst's speciality was the artillery, his lectures, like his handbook, were full of geometric tables and calculations of best angles of attack, and statistics about the penetrativity of missiles at particular distances, or the poor quality of the British-*cum-*

Hanoverian cannon. Other than that, he followed the Vegetian pattern religiously.

Scharnhorst and all these other authors wrote with a readership of officers in mind, their manuals being devised as a course book for officers in training, and as a work of reference 'for use in the field', when one grim, rainy morning the officer might wake up to see the need to cross a flooded riverbed, move camp, or launch a surprise attack on an advancing enemy, and wanted to collect his thoughts, with the help of such a book, to remember to make all necessary provisions and to proceed in the most sensible way.

Clausewitz by contrast had a totally different aim when writing *On War*, an aim which no author before him had had in quite the same way. He did *not* want to write primarily about *how* to wage war, although *On War* contains some unoriginal 'books' or chapters dealing with many of these subjects as well (Books 3–4, see below). Instead, he wanted mainly to explore the phenomenon of war, in its tangible, physical, and psychological manifestations. He wanted to analyse war, to understand it better. He contrasted his own aim in writing *On War* with that of the authors who had gone before him: Clausewitz wrote of his own theories that they were 'meant to educate the mind of the future commander, or more accurately, to guide him in his self-education, not to accompany him to the battlefield; just as a wise teacher guides and stimulates a young man's intellectual development, but is careful not to lead him by the hand for the rest of his life'. He explicitly dismissed manuals as pointless, as theoretical rules could not possibly apply to all real cases: 'a positive doctrine', he wrote, 'is unattainable' (pp. 89 f.).

Clausewitz assumed that understanding the essence, the nature (*Wesen*) of war would eventually help future leaders wage and win their wars more effectively, and this, to him, was the ultimate aim of the exercise. He has therefore been admired by strategists and leaders who sought to win wars more effectively, more decisively, and faster, regardless of whether their motives in waging war have been judged good or bad by history. The Americans, Germans, French, British, Russians—Imperial and Communist—and the Chinese under Mao studied *On War* in search of useful lessons, to be applied in wars seen retrospectively in many different guises: as just wars or imperialist wars, wars to impose German nationalist and racist aims upon the world, war *against* Nazi Germany, wars of liberation, or wars of colonial expansion. For Stalin, Clausewitz was the symbol of the strategy of German fascism and was therefore dismissed as not only bad but also useless, as Stalin had defeated German fascism. For many writers of the early Cold War, Clausewitz's dictum that war was the continuation of politics by military means was inapplicable because irrational in the nuclear age, and again he was dismissed. Later on, when it became clear that the Cold War was not only characterized by the threat of war—major war in Europe, or major war elsewhere between the superpowers—but also formed the backdrop to many actual wars outside Europe, Clausewitz was resurrected and once again achieved major prominence among those who sought to learn how best to wage war.[1]

Nevertheless, his approach of studying the phenomenon of war in order better to understand it can equally serve anybody aiming to limit or even eliminate war from the world, from pacificists to pacifists and peace researchers.[2] Clausewitz himself never expressed any doubt that war was an eternal human social phenomenon. In Book 4 he wrote,

We are not interested in generals who win victories without bloodshed. The fact that slaughter is a horrifying spectacle must make us take war more seriously, but not provide an excuse for gradually blunting our swords in the name of humanity. Sooner or later someone will come along with a sharp sword and hack off our arms.[3]

It is a useful reminder that this pessimism is gleaned from the experience of Clausewitz's own lifetime, when it was—for once—not Prussia which initiated war with France, but outside powers that brought war to Prussia. Developments since his lifetime, and the further growth of international law and international organizations which are a source of hope to those who, like Kant and the fathers of the UN Charter, dream of a world in which war is outlawed and only employed by entities excluded from the community of nations, could not easily be foreseen by Clausewitz. This fact should not, however,

---

[1] Beatrice Heuser, *Reading Clausewitz* (London: Pimlico, 2002).

[2] For the difference between pacificists (who prefer peaceful settlements of disputes while acknowledging the possibility of a just war) and pacifists (who absolutely reject all war as evil), see Martin Ceadel, *Pacifism in Britain, 1914–1943: The Defining of a Faith* (Oxford: Clarendon, 1980).

[3] *On War* Book 4, Chapter 11, p. 260 of the Princeton Text, passage omitted in this edition; see also *On War* Book 1, Chapter 2, p. 98.

detract from the usefulness of his approach to those who seek to understand war in order to overcome it.

It is in Clausewitz's discussion of the human and social factors of war, not the tactical or technological ones, that one can find the greatest *lasting* wisdom of Clausewitz's observations and analysis, and to which we will turn below. But first we need to consider the education and intellectual approach which led Clausewitz to achieve this Copernican leap in our thinking about war.

Clausewitz's background and education

Perhaps, in good German Protestant tradition, Carl von Clausewitz owed his intellectual brilliance to a lineage of pastors, particularly to his grandfather Benedictus Gottlob Clauswitz [*sic*], pastor in Saxony and later professor of theology at the University of Halle. But if Carl inherited his grandfather's cleverness, he probably owed little to the intellectual environment provided by his parents. The household into which Carl Phillip Gottlieb was born on 1 June 1780 in a provincial town in Brandenburg, as the fifth of six children, was anything but intellectual. His mother, Friederike Dorothea Charlotte Schmidt, was the daughter of a local civil servant. His father Friedrich Gabriel, the youngest of six brothers, had been only 9 years old when his father, the professor, died, and had joined the armed forces with the help of his stepfather, a Prussian major, to obtain a regular income. Previously, there had been no military tradition in the family. Friedrich Gabriel, who never rose beyond the rank of a Prussian lieutenant, had been badly wounded in the Russian siege of Colberg (1758–61), and, as a veteran, had been given administrative duties in the small town of Burg near Magdeburg. In his household there was little left, it seems, of the piety of his great-grandfather Johann Carl Clauswitz [*sic*], also a pastor in Saxony. Carl von Clausewitz in his writing made no references to religious questions, Christian-inspired morality, or indeed God (other than—blasphemously, as his great-grandfather would no doubt have thought—describing Napoleon as the 'God of War'). Out of modesty, or because it was never properly confirmed, the great-grandfather and the grandfather had not used the little 'von' in front of their name that denotes nobility, while Carl's father had asked for royal permission to use it again, granted to the veteran soldier. Nevertheless, Carl's father could neither afford nor was he inclined to show much of the lifestyle associated with the noble classes at the time. Instead, if there were guests in the unspectacular townhouse of the Clausewitzes in Burg, they were mainly old comrades of Carl's father, of a rough and ready sort, as Carl later admitted to his fiancée, Countess Marie von Brühl, who was classes above him socially (an obstacle to their union which took them years to overcome).[4]

Young Carl and his siblings would have received a decent primary school education in his native Burg, up to the age of 11 in Carl's case. He himself confided to his future wife that the education he received there was 'pretty mediocre'.[5] While it has been claimed that Clausewitz learnt Latin at this primary school,[6] this seems unlikely, as he rarely if ever used Latin words, let alone quotations, and showed a pronounced lack of interest in the wars of Antiquity. Nor did Clausewitz attend any grammar school, the traditional place to become acquainted with classical languages: at the age of 11, Carl, like his older brothers Friedrich and Wilhelm before him, became cadets ('Junkers') in Prussia's army (only the oldest, Gustav, stayed out of the army and became a tax inspector). All three would rise to the rank of general, Friedrich and Wilhelm decorated with the order of *pour le mérit*. Carl spent his teens in active military service, including both a campaign and garrison duty in Neuruppin just to the north of Berlin. He said himself that he read more than others during this time, and scholars have subsequently speculated about the availability of books to him in the library of one of the Hohenzollern princes, who had his residence nearby.[7]

At some stage Carl must have learnt French, which was still the language in which people communicated much of the time at court. As Clausewitz's widow later recalled, they exchanged polite niceties in French when they first met.[8] This is of relevance, because Clausewitz could read and was clearly influenced by French literature

---

[4] 'News of Prussia in its great catastrophe', in Eberhard Kessel (ed.), *Carl von Clausewitz: Strategie aus dem Jahr 1804 mit Zusätzen von 1808 und 1809* (Hamburg: Hanseatische Verlagsanstalt, 1943), 9.

[5] Karl Linnebach, *Karl und Marie von Clausewitz: Ein Lebensbild in Briefen und Tagebuchblaettern* (Berlin: Martin Warneck, 1917), 83.

[6] Karl Schwartz, *Leben des Generals Carl von Clausewitz und der Frau Marie von Clausewitz geb. Gräfin Brühl*, vol. i (Berlin: Ferd. Dümmler Verlag, 1878), 33.

[7] Peter Paret, 'The Genesis of *On War*', in *Carl von Clausewitz: On War*, ed. and trans. Michael Howard and Peter Paret (Princeton: Princeton University Press, 1984), 7.

[8] Linnebach, *Karl und Marie von Clausewitz*, 40.

on war; other important texts were accessible for him only in German translations, where these existed.

Between 1795 and 1801, Clausewitz would have attended some classes, as regiments were obliged to develop further the skills of their young officers not only in practical exercise but also in the classroom. The education of officers, particularly of those destined for higher ranks, had developed a very distinct mathematical, scientific dimension since the time of the 'Military Revolution' in the sixteenth century, when tactics and drill adapted to the new firepower, particularly to hand-held weapons.[9] The introduction of cannon, and of firearms that could be held and operated by one man, increased the need for calculations of flight paths of missiles, and of the most useful deployment of artillery on the battlefield or in the siege of a fortified place. When the Italian marquess Annibale Porroni was writing in the late seventeenth century, the standard education of a future general included: geometry, arithmetic, trigonometry, and also the measuring of spaces which Porroni subsumed under the terms stereometrics, logimetrics, planimetrics, and topography. Important in this context was the art of drawing maps and sketches, which Porroni called iconography. A future general's education would also include basic mechanics, hydraulics, geography, and geodesics, and what Porroni called hydrography and nautical skills.[10] Tactics were also taught, and Porroni's *Universal Modern Military Treaty*, like many other works on the art of war and generalship at the time, supplied the military commander with all the extra knowledge he needed concerning such things as the movement of troop units, how to discipline them before and during battle, how to lay sieges, build fortresses, choose officers for different ranks and duties, deploy cannon and siege engines, and so on.

The basic education of cadets had not changed much by the time Clausewitz came to Neuruppin. We can make some inferences about it from the writings of Clausewitz's future mentor, Gerhard von Scharnhorst. Scharnhorst wanted it to consist of neat handwriting and orthography (something the Germans are sticklers for to the present day), and good style in writing. These subjects were seen as

---

[9] See Michael Roberts, *The Military Revolution, 1560–1660* (Belfast: Marjory Boyd, 1956).

[10] Marchese Annibale Porroni, *Trattato universale militare moderno* (Venice: Francesco Nicolini, 1676), 403–6.

just as important as the other subjects of study, arithmetic and geometry, geography (or in actuality, map reading) with almost exclusive reference to Europe, logic, 'war sciences' and 'field work', i.e. practical exercises in the field. Arithmetic, geometry, and particularly trigonometry were extremely important for the young officers dealing with artillery, or logistics. Beyond this, the level of teaching in the regiments must generally have been fairly basic.[11]

It was only at the age of 21, in 1801, that Carl received further formal education upon entering Lieutenant Colonel Scharnhorst's Academy. Here Clausewitz had a little more than two years of training, with the same spread of subjects which Scharnhorst had set down for the regimental education: maths, geography, 'war sciences', and practical exercises in the field. Scharnhorst wanted all that was taught and learnt to have practical applicability. He thought that too much maths was being taught, as his pupils were not destined to become engineers.[12] Clausewitz does not seem to have shared Scharnhorst's limited enthusiasm for higher maths. Indeed, Scharnhorst's record of the performance of his students at the Academy for Young Officers, written towards the end of Clausewitz's time there, in early 1804, tells us that Clausewitz was particularly good at maths and 'war sciences', while gifted with good judgement and a good presentational style.[13]

Beyond arithmetic and geometry, what might Clausewitz have picked up? Somewhere along the way he must have encountered some physical experiments, involving particularly electricity and magnetism and their effects. He clearly knew of and admired Newton, whose great theorems must have been in his mind when trying to formulate theories on war (p. 59). In emulation of Newton's discoveries in physics, Clausewitz sought to find the laws that govern war (pp. 101 ff.). He must also have picked up some higher maths, as he writes about 'co-efficients' and factors influencing strategy. Clausewitz clearly knew the work of Leonard Euler, who from 1741 to 1766 had been a member of the Prussian Academy of Sciences in Berlin, and whom Clausewitz clearly admired (p. 59). Euler had coined the term 'function', and invented the concept of one thing being a function of another, simply expressed in the formula $y = f(x)$,

[11] Gerhard von Scharnhorst, *Private und Dienstliche Schriften*, vol. iii: *Preussen 1801–1804: Lehrer, Artillerist, Wegbereiter* (Cologne: Boehlau Verlag, 2005), 111–26.
[12] Ibid. 88–90.   [13] Ibid. 150.

where y is a function of x. By the time Clausewitz pursued his studies, the idea of functions with several interdependent variables was not yet being taught in classrooms, but the logical step could be easily taken, and *was* taken by Clausewitz, as I shall argue shortly.

But first, it is worth explaining what was meant by 'war sciences' at the time.[14] Classes here revolved around two things: lectures on the history of wars, and especially on the history of campaigns and battles, and the re-enactment of elements of these campaigns, either with maps and drawings, or on an open field somewhere in place of the actual historical battlefield of whichever campaign was being studied. Such military history during Clausewitz's time at the Academy focused on the Wars in the Netherlands from 1745 to 1748 and 1793 to 1794; the Seven Years War; and the French Revolutionary Wars. These would also be the historical wars on which Clausewitz later drew himself, adding only the campaigns of Gustavus Adolphus in the Thirty Years War, on which he wrote a research paper of his own.[15]

How were these wars studied? We have Scharnhorst's own syllabus to tell us that the tutor would begin by discussing with the students what the statistical (demographic, economic) and military situations of the belligerent states were, i.e. the size, quality, equipment, and disposition of their armed forces. Then classes would focus on descriptions of the natural and military characteristics of each theatre of war. The students would derive all this from maps and written sources, being encouraged to use sources critically, as history students might some decades later when history became an established field of studies at Germany's universities. The same critical study would apply to the accounts of the campaigns themselves, written by historians and chroniclers. From these the students were to establish their own illustrated accounts of the campaigns. There is nothing to lead us to believe that political and ideological aspects of the wars, legal aspects, or overall effects on the countries involved were touched upon in any way.[16]

---

[14] Ibid. 336–40 has a lecture on this subject by Scharnhorst, given presumably during Clausewitz's time at the Academy.

[15] 'Verfassung und Lehreinrichtung der Akademie für junge Officiere und des Instituts für die Berlinische Inspektion', in *Gerhard von Scharnhorst: Ausgewählte Militärische Schriften*, ed. Hansjürgen Usczeck and Christa Gudzent (East Berlin: Militärverlag der DDR, 1986), 208 f.

[16] Ibid. 209.

War as an instrument of state policy

It is safe to say that nothing in his education or the writing of his teacher, Scharnhorst, presaged the intellectual quantum leap which Clausewitz made in emphasizing the link between the political context and the resulting *aims* of the belligerents and war. The stimulation to take this logical step must have come from elsewhere, but Clausewitz was not the only one to make it. Perhaps he gained more from the exchange with his fellow-students than he ever admitted. Clausewitz's classmate in the Academy and later fellow-teacher at the War School in Berlin, Johann Jakob Otto August Rühle von Lilienstern (1780–1847), was the first to spell out this link between politics and war in his revision of Scharnhorst's Field Manual published in 1817/18 (which otherwise followed Scharnhorst's structure in the classical Vegetian tradition described above).[17] Rühle wrote:

There is a Why? and a What For?, a purpose and a cause, at the bottom of every war and every [military] operation. These will determine the character and the direction of all activity.

The individual operations have military purposes; the war as a whole always has a final political purpose, that means that war is undertaken and conducted in order to realise the political purpose upon which the State's [leading] powers have decided in view of the nation's internal and external conditions.[18]

Famously, Clausewitz turned this around to read that war is the continuation of politics (pp. 20–1, 27–9). In the case of Clausewitz, the understanding of the nexus between politics and war was due not only to the teaching of Scharnhorst and other teachers in the Academy. Clausewitz also read and greatly admired Machiavelli's *The Prince*[19] in German translation, where war is one of many tools the prince uses for his political ends, and the work of an outstanding French author on war, who like Clausewitz had broken the mould of the Vegetian tradition of writing on the subject. This was Count Jacques Antoine Hippolyte de Guibert (1743–90), whose work the

[17] R[ühle] von L[ilienstern], *Handbuch für den Offizier zur Belehrung im Frieden und zum Gebrauch im Felde*, vol. i (Berlin: G. Reimer, 1817).

[18] Ibid., vol. ii (Berlin: G. Reimer, 1818), 8.

[19] *Carl von Clausewitz: Verstreute kleine Schriften*, ed. Werner Hahlweg (Osnabrück: Biblio Verlag, 1979), 157–66 contains Clausewitz's letter to the philosopher Fichte on the subject.

intellectually more pedestrian Scharnhorst used in his lectures (if only to illustrate a point about his beloved artillery).[20]

Clausewitz (and perhaps Scharnhorst) owed to Guibert a very important interpretation (or 'narrative', as one would say today) of the contrast of the wars of the Ancien Régime and the wars of the new era in which they were living. Guibert, scion of the age of Enlightenment, was the foremost French thinker on military affairs among the *Lumières* and *Encyclopédistes*, those great thinkers, many of whom participated in the creation of the first French Encyclopedia.[21] In a gripping passage in the brilliant work of his youth, the *General Essay on Tactics* (which was in fact a treatise on most aspects of war), he had already breached the chasm that existed elsewhere between the war manuals on the one hand, and, on the other, legal or political-philosophical writings on war in the style of Machiavelli's *Prince* and *Discourses on Livy*, the writings on the law of war of Justus Lipsius and Hugo Grotius, or the political philosophy of Jean-Jacques Rousseau. As a young officer Guibert had experienced and smarted under France's poor performance in the Seven Years War, the war which on the Prussian side, even in Clausewitz's school days, was seen as the supreme example of how to wage and win a war. From his experience, Guibert had developed a great admiration for the Prussian way of war, and yet he felt that it could be topped, at least in theory. Guibert described eighteenth-century Europe as full of

tyrannical, ignorant or weak governments; the strengths of nations stifled by their vices; individual interests prevailing over the public good [common wealth]; morals, that supplement of laws which is so often more effective than them, neglected or corrupted; . . . the expenses of governments greater than their income; taxes higher than the means of those who have to pay them; the population scattered and sparse; the most important skills neglected for the sake of frivolous arts; luxury blindly undermining all states; and governments finally indifferent to the fates of the people, and the peoples, in return, indifferent to the successes of governments.[22]

---

[20] Scharnhorst, *Private und Dienstliche Schriften*, iii. 224.

[21] The great thinkers of the Age of Enlightenment, many of whom participated in the project to distil all knowledge of the world in the first great French Encyclopedia.

[22] Guibert, 'Essay général de tactique', in Guibert, *Stratégiques* (Paris: L'Herne, 1977), 135. Here and in the following, my translation, DBGH.

Sarcastically, he noted that the effect was that Europe seemed 'civilized'.

Wars have become less cruel. Outside combat, blood is no longer shed. Towns are no longer destroyed. The countryside is no longer ravaged. The vanquished people are only asked to pay some form of tribute, often less exacting than the taxes that they pay to their sovereign. Spared by their conqueror, their fate does not become worse [after a defeat]. All the States of Europe govern themselves, more or less, according to the same laws and according to the same principles. As a result, necessarily, the nations take less interest in wars. The quarrel, whatever it is, isn't theirs. They regard it simply as that of the government. Therefore, the support for this quarrel is left to mercenaries, and the military is regarded as a cumbersome group of people and cannot count itself among the other groups within society. As a result, patriotism is extinct, and bravery is weakening as if by an epidemic.[23]

This is of course a difficult argument to follow for those who hope to eliminate war altogether, and who welcome limitations on war, especially the sparing of non-combatants.[24] But in his youthful fervour, Guibert the soldier obviously smarted from the indifference of the French population as a whole to the efforts and suffering of the French Army, which, in his view, had led to France's defeats and Prussia's success in the Seven Years War.

'Today', continued Guibert,

the States have neither treasure, nor a population surplus. Their expenditure in peace is already beyond their income. Still, they wage war against each other. One goes to war with armies which one can neither [afford to] recruit, nor pay. Victor or vanquished, both are almost equally exhausted [at the end of a war]. The mass of the national debt increases. Credit decreases. Money is lacking. The fleets do not find sailors, armies lack soldiers. The ministers, on one side and on the other, feel that it is time to negotiate. Peace is concluded. Some colonies or provinces change hands. Often the source of the quarrels has not dried up, and each side sits on the rubble, busy paying his debts and keeping his armies alert.

---

[23] Ibid. 187 f.

[24] It is, incidentally, a matter of debate how 'humane' warfare was in Guibert's own time—recent historiography suggests that the wars of the Ancien Régime had drastic consequences also for non-combatants, in the shape of famine and starvation. The wars in North America in the eighteenth century, moreover, had pronounced genocidal elements. See Stig Förster and Roger Chickering (eds.), *War in an Age of Revolution: The Wars of American Independence and French Revolution, 1775–1815* (expected 2006).

But imagine [he continued] that a people will arise in Europe that combines the virtues of austerity and a national militia with a fixed plan for expansion, that it does not lose sight of this system, that, knowing how to make war at little expense and to live off its victories, it would not be forced to put down its arms for reasons of economy. One would see that people subjugate its neighbours, and overthrow our weak constitutions, just as the fierce north wind bends the slender reeds. . . . Between these peoples, whose quarrels are perpetuated by their weakness [to fight them to the finish], one day there might still be more decisive wars, which will shake up empires.[25]

It is impossible to read this passage without thinking of the *levée en masse* (the massive recruiting of volunteers for the French Army, and the mobilization of the population) under the French Revolution, and of the Napoleonic achievements. Napoleon was clearly the *aquilon*, the fierce north wind, which swept across Europe bending the slender reeds of the old monarchies. And this is precisely the idea that occurred to Scharnhorst,[26] Clausewitz, and many contemporaries. This passage was translated almost verbatim into German by Clausewitz in a paper he wrote in 1812,[27] which is why it is quoted here at such length, and indeed Clausewitz paraphrased and elaborated on it in Book 8 Chapter 3B of *On War*, taking the narrative further, in the light of the French Revolutionary and Napoleonic Wars he had experienced:

This was the state of affairs at the outbreak of the French Revolution. . . . in 1793 a force appeared that beggared all imagination. Suddenly war again became the business of the people—a people of thirty millions, all of whom considered themselves to be citizens. . . . The people became a participant in war; instead of governments and armies as heretofore, the full weight of the nation was thrown into the balance. The resources and efforts now available for use surpassed all conventional limits; nothing now impeded the vigour with which war could be waged . . . (pp. 234–8)

Crucial to this transformation, of course, were the values of the French Revolution and the confidence which the feeling of defending one's own cause as a citizen instilled in France's revolutionary armies. But as a counter-revolutionary and hater of all things

[25] Guibert: 'Essay général de tactique', 137 f.

[26] Scharnhorst, *Private und Dienstliche Schriften*, iii. 71.

[27] *Carl von Clausewitz: Schriften—Aufsätze—Studien—Briefe*, ed. Werner Hahlweg (Göttingen: Vandenhoek und Ruprecht, 1966–), i. 710 f.

French, Clausewitz wrote little (and nothing else in *On War*) about the contesting political ideologies of his times, or about how these might influence the political aims of warfare.[28] Clausewitz only noted that the aims pursued by a party in waging war might vary considerably from very limited—the conquest of a hamlet, perhaps, with the mere purpose of using it as a bargaining chip in peace negotiations—to very extensive—the conquest of a large country. Clausewitz was curiously uninterested in exploring, in *On War*, how ideology determined the extent or limitations of war aims, even though he noted, in Book 8, that every age, every culture had had its own style of waging war, and that war aims differed accordingly. But by having spelled out the nexus between state politics and war, Clausewitz alerted generations of scholars and analysts to this crucial interface, laying the foundation, one might say, for future strategic or security policy studies.

So much for the intellectual background that Clausewitz acquired in his formal education. He lived up to his own maxim, however, that the officer should study under his own guidance and discipline, which included studying the world around him, and constantly enlarging his 'data base' as we might now say, his collection of relevant case studies, from which to draw conclusions about the essence of war. Some narrow-minded historians today might disparagingly call Clausewitz a political scientist for espousing this methodology. But this Clausewitzian methodology—the deduction of theory from a multitude of historical examples—was among the most conventional aspects of Clausewitz's work. It had been used by the field manualists since Vegetius, albeit with the aim of deriving firm rules of conduct, not of gaining a better understanding of the phenomenon of war as such.

Not only did Clausewitz study historical cases of wars (concentrating, as we have noted, on the time since the Thirty Years War, i.e. mainly on wars between sovereign states[29]), although these formed the basis of his collection of data. He also analysed the wars of his own times, some of which he had experienced at close quarters, as an astute observer and analyst.

---

[28] *On War*, Book 8, Chapter 3, pp. 227 f.

[29] Which has led some to argue that Clausewitz has nothing to say about wars not waged between sovereign states—an untenable argument, as we shall see.

Clausewitz's own experience

Clausewitz took part in four military campaigns. As a 12- and 13-year-old, Carl experienced the War of the First Coalition against France, which took him to the Rhineland; he witnessed the burning of Mainz, which, although signifying its liberation from the French, meant a large-scale destruction of the beautiful city. As a young lad, Clausewitz did not appreciate the implications, and as he later shamefacedly admitted to his fiancée, he cheered along with the other soldiers to see Mainz go up in flames.[30] The next campaign he witnessed was that of 1806, when Prussia was defeated at the hands of Napoleon at the double battle of Jena and Auerstedt on 6 October. Clausewitz was at Auerstedt as aide de camp to the Prussian prince Augustus Ferdinand, who headed his regiment. Augustus Ferdinand refused to admit defeat, and his regiment together with some others retreated to the north of Brandenburg, where they were routed at the small battle of Prenzlau on 28 October 1806. Augustus Ferdinand was taken prisoner together with Clausewitz, and held in France until both were released in the autumn of 1807.

Meanwhile the Prussian court had left French-occupied Berlin and had moved to Eastern Prussia, residing alternately in Königsberg and Memel until 1809, when it returned to Berlin. Clausewitz's old patron, Scharnhorst, invited Clausewitz to join him there, which he did. In Königsberg and Memel, Clausewitz participated in the work of a commission under the generals Scharnhorst and Gneisenau which reformed the structure of the Prussian military and indeed the state as a whole. While Clausewitz was nowhere near any field of battle for the next five years, he was highly politicized, moving in military circles that loathed the French and were consequently highly critical of their king Frederick William III for behaving so accommodatingly towards Napoleon in the Franco-Prussian Peace Treaty of Tilsit of 1807. With admiration, Clausewitz watched from afar the Tyrolean insurrection of 1809 against the French occupying forces, on the basis of which he later developed his own elaborate policy plans on how Prussia's peasant population should arise in a *Landsturm* or popular uprising against Napoleon.[31] By contrast, we

[30] Linnebach, *Karl und Marie von Clausewitz*, 83.

[31] Clausewitz's 'Bekenntnisdenkschrift' of February 1812, in *Carl von Clausewitz: Schriften*, ed. Hahlweg, i. 682–750, here esp. p. 721.

have no evidence that he followed or took an interest in the Spanish resistance against the French in their famous *guerrilla* (small war) of 1808–12.

From 1810 Clausewitz secured a teaching position at the General Military School in Berlin, the successor-institution to Scharnhorst's Academy. Despite the difference in social standing, he was finally allowed to marry Marie Countess von Brühl in 1810, to whose position at court he probably owed the honour of becoming private tutor to the crown prince Frederick William (later IV) from 1810 to 1812. He was also promoted to the rank of a major.

Resisting Napoleon, 1812–1815

Clausewitz remained in close contact with Scharnhorst and his new mentor from the circle of reformers, General August Neidhardt von Gneisenau. They all baulked at King Frederick William III's continued toeing of the French line: when Napoleon prepared to invade Russia in February 1812, Frederick William signed a pact of alliance with Napoleon against the country, Prussia's eastern neighbour. Along with a number of like-minded Prussian officers, Carl von Clausewitz resigned in protest. While his brothers Frederick and William dutifully continued to fight for Prussia and against Russia, Carl offered his services to the very state that was still forcefully opposing the French: imperial Russia itself. The next series of campaigns in which Clausewitz participated from 1812 to 1815 provided him with most material for his analysis of war.

Clausewitz knew no Russian, but this was also true for many other officers in Tsar Alexander I's services, and the higher ranking Russian officers all spoke French. The linguistic barrier thus did not prevent Clausewitz from writing a remarkably detailed account of the campaign of 1812 based on his own memoirs and on further studies undertaken by him in 1823.[32] He witnessed the battles of Smolensk and Borodino on the Moskva River, and the French seizure of Moscow. According to Clausewitz, it was said in the Russian camp at the time that as the Russians were retreating, Moscow caught fire by accident and not by intention, leading to its famous destruction and the death of many Russian civilians and wounded

[32] *Carl von Clausewitz: Schriften—Aufsätze—Studien—Briefe*, ed. Hahlweg, ii (1990), pt. 2, 717–924.

soldiers.[33] Clausewitz was particularly impressed by the constant harassment of the retreating French army by peasants armed with anything from pitchforks to muskets. From this he drew important lessons about the 'arming of the people' and the 'people's war' (Book 6, Chapter 26). He was also greatly impressed by Russia's capacity to ride out the storm on account of the country's size, the width of its rivers—the Beresina would form a particularly grim obstacle—the poor quality of its roads, the ravages of winter, and also the determination of the Russian people to hold out. These factors taken together—geography, climate, battles, and guerrilla warfare on the advancing and then retreating forces—reduced Napoleon's armies of almost 600,000 at the beginning of the campaign to a mere 20,000 at its end, when there were no more than 180,000 Russian troops facing them.[34] The impression of this stunning defeat of the 'God of War' and the utter destruction and tragic wastage of his armies, confronted with a much smaller army acting in self-defence, left its indelible mark on Clausewitz's thinking, as we can see in Book 6 of *On War*. Here Clausewitz extolled the superiority of a defensive strategy over the offensive, much to the annoyance of subsequent generations in Germany, France, America, and elsewhere, who in the late nineteenth and early twentieth enturies for ideological reasons much preferred *l'offensive à l'outrance*, the offensive at all costs, as a sign of vigour, initiative, and national prowess. Indeed, until the 1812 campaign, Clausewitz himself had thought a defensive war very regrettable.[35]

It was also in this context that Clausewitz developed his theory of diminishing returns. He extrapolated from Napoleon's war against Russia that the attacker had all the impetus and the élan on his side, but that by and by he would run out of steam, particularly when invading an almost limitless space, with a population determined to hold out against the invader, who could retreat into the interior of the country.[36] Even a victory in battle in such a campaign, which at first might seem highly advantageous to the side of the invading

[33] Ibid. 869.

[34] Charles J. Esdaile, *The Wars of Napoleon* (London: Longman, 1995), 252–61, which shows the impressive accuracy of Clausewitz's calculations of 1823.

[35] Clausewitz, 'Strategie aus dem Jahre 1804', in *Clausewitz: Verstreute Kleine Schriften*, ed. Hahlweg, 25.

[36] *On War*, Book 6, Chapter 25, a section omitted in this edition.

forces, might represent the culmination or turning point of the attack, and mark a decline in the attacker's forces and stamina, only to lead to his eventual defeat. All this was very well illustrated by Napoleon's ill-fated campaign against Russia (Book 7, Chapter 22).

As this campaign was reaching its nadir, at the end of 1812, Clausewitz was chosen as intermediary between the Russian and the Prussian general staffs for negotiations that were held at Tauroggen. Leading the negotiations on the Prussian side, Ludwig Count Yorck von Wartenburg decided to end Prussia's alliance with France, forcing the hand of the Prussian king: the famous neutrality treaty of Tauroggen between Russia and Prussia was signed on 30 December 1812, followed on 28 February 1813 by a treaty of alliance between Russia and Prussia and on 16 March 1813 by a declaration of war by Prussia against France.

Here began the Prussian 'wars of liberation' (from French oppression, as Clausewitz and his friends saw it), in which Clausewitz participated actively, first on the Russian side, then back in Prussian service. In 1813, Clausewitz joined the Prussian army of Wittgenstein in East Prussia, which rose up in arms against the French, Clausewitz actively helping to organize this popular insurrection from Königsberg. Clausewitz was commissioned by Scharnhorst together with Count Alexander Dohna to work out ways of arming the population, and of integrating all aspects of the 'people's war' and irregular warfare ('small wars') into their resistance against the French.[37] Clausewitz was with Wittgenstein's army when Berlin was liberated from French occupation in March 1813, but shortly afterwards the Prussian armies were on the move again, and on 2 May and 20 May 1813, Clausewitz was in the thick of two battles against Napoleon at Großgörschen and Bautzen, both won by the Prussians. (Scharnhorst was wounded at Großgörschen and shortly after died of his wounds.) In the following months, Clausewitz was involved in negotiations to bring further powers, especially Denmark and Sweden in the north, alongside Prussia in the big counter-attack on Napoleonic France. He joined the Army corps of Count Louis George Thedel of Wallmoden-Gimborn as his quartermaster-general. Wallmoden's forces were formally part of the Swedish Crown Prince's Northern

---

[37] Schwartz, *Leben des Generals von Clausewitz*, vol. ii (Berlin: Ferd. Dümmler, 1878), 5–8.

Army, and in 1813 it joined the Prussian forces in its campaigns against Napoleon.[38]

Clausewitz was only properly readmitted into the Prussian army on 11 April 1814 as Colonel of the Infantry, joining the General Staff a year later and becoming chief of the general staff of Lieutenant General von Thielmann's Third Army Corps (which in turn stood under Blücher's general command). From March 1815 Clausewitz was fully involved in the great campaign to counter Napoleon's return from Elba and his attempt to turn back the wheel of history.

In June 1815 Clausewitz fought in the battle of Ligny, and on 18 June was with Thielmann's forces against the French under Marshal Grouchy at Wavre, while Napoleon, Wellington and Blücher clashed at nearby Belle Alliance (Waterloo). The hostilities at Wavre outlasted the victory at Waterloo, and finally Grouchy managed to get away, and with his troops reached Paris before the Prussians did. This earned Thielmann considerable criticism; he in turn blamed his chief of staff, Carl von Clausewitz, which was to have an adverse effect on his subsequent career.[39]

The return of limited wars

Militarily, the period of 1812 to 1815 was the peak of Clausewitz's career. When he returned to Berlin in 1815 with the victorious army of Blücher, he found that the king had not forgiven him for his act of treason in 1812, the escape of Grouchy cast a shadow on his reputation, and he was punished—or so he felt—by being given the job of an administrative director of the General War School in Berlin, albeit at the rank of a general. These were the years in which he reached his intellectual zenith, however, as it was in the years between 1815 and 1830 that he wrote *On War*.

In 1830 he was called back to active duty. He was appointed Inspector of the second Artillery Group in Wrocław (Breslau), that had formerly been part of the Kingdom of Poland and was populated in large part by ethnic Poles. In late 1830 and early 1831, a Polish insurrection against the Russian rule centring on Warsaw to the east threatened to spread to these Prussian-held territories. Clausewitz, together with his old mentor, now Field Marshal, Gneisenau, made a

[38] Schwartz, *Leben*, ii. 30–2.   [39] Ibid. 129–34.

number of inspection tours taking them, *inter alia*, to Poznan (Posen), but without seeing any serious combat; the insurrection was quelled by the Russians alone and did not spread any further. And yet, other dangers awaited the Prussians. When in Wrocław, Clausewitz and Gneisenau both succumbed to a cholera epidemic which vastly decimated the occupation forces. Clausewitz died on 16 November 1831.

This last campaign came too late to influence *On War*, but Clausewitz still had time to write reflections about the events in the Russian-held parts of Poland, and the other political events in Europe from Belgium and the Netherlands to northern Italy, in his diary and in letters to his wife.[40] The occurrence of more limited wars than those of the French Revolution and the Napoleonic campaigns, such as the series of wars in which Russia engaged from the mid-1820s (and which gave the Poles the hope of being able to assert themselves against a weakened Russian Empire), clearly set him thinking. In 1827 he noted the need to revise existing 'books' or sections of *On War* in a very major way (p. 7 f.). Until now—and he had just about written Books 1 to 6—he had concentrated on the French Revolutionary and Napoleonic wars in his analysis of war. He now felt that this had been an excessive limitation of his focus. Hitherto, he had more or less assumed that war would develop in a linear fashion, from the more limited wars of the Ancien Régime described by Guibert to the unbounded raging of war under Napoleon prophesied by Guibert and witnessed by Clausewitz in his own lifetime. In 1827, however, he realized that the return to more limited forms of war (which could clearly be observed in Russia's campaigns against the Ottoman Empire) implied a fluctuation in the forms of war that really manifested themselves in the world, and that war could come in all different shapes and forms.[41] This consequently meant that he needed to rethink all he had written in his manuscript, treating the recent great wars against France and the particular French ways of war of 1792 to 1815 as only one among many paradigms. This called for major revisions in his text.

The last two books that Clausewitz wrote, Books 7 and 8, reflect this new realization that war was manifold in its real manifestation. Clausewitz also revised the existing scripts of Books 1 and 2. But

[40] Ibid. 298–401.
[41] *On War*, Book 8, Chapter 3, pp. 230–40 and Chapter 6, p. 256 f.

before completing his revisions, he was mobilized, and his early death at the age of 51 prevented him from making the changes that his new understanding should have necessitated. This accounts for a host of contradictions in *On War*, mainly in Books 1–6, where war is often described exclusively in the light of the French Wars of 1792–1815. In these books, the be-all and end-all of war is victory in battle, and we typically find passages such as this:

What do we mean by the defeat of the enemy? Simply the destruction of his forces, whether by death, injury, or any other means—either completely or enough to make him stop fighting. . . . the complete or partial destruction of the enemy must be regarded as the sole object of all engagements.[42]

By contrast, elsewhere, especially in Books 1 and 8, the manifold forms of war, the 'true chameleon', are duly taken into account in the analysis, and a much subtler approach is taken: physical destruction is not the ultimate aim, but a psychological victory is.

Clausewitz's main theories

To appreciate the main conclusions that Clausewitz drew from his studies of all these wars, we should first begin with his main definition of war, which has become a particularly helpful key to our understanding of the subject. In Book 1, we read:

War is nothing but a duel on a larger scale. Countless duels go to make up war, but a picture of it as a whole can be formed by imagining a pair of wrestlers. Each tries through physical force to compel the other to do his will; his *immediate* aim is to *throw* his opponent in order to make him incapable of further resistance.
   War is thus an act of force to compel our enemy to do our will. (p. 13)

While Clausewitz does not spell this out, we can from this deduce that success in war means imposing one's will upon the enemy and 'persuading' him through the use of force to desist from pursuing his opposed aims. This intelligent definition of success in war, or victory, thus does not score up the enemy's dead or the winning of one battle, but defines victory as the achievement of one's own war aims

---

[42] Ibid., Book 4, Chapter 3, p. 227 (Princeton edn.)—this passage has been omitted in this edition.

and the imposition of one's will upon the enemy. This is an extremely helpful test of the success of any military campaign.

It also becomes immediately clear that victory, defined this way, does not necessarily have to culminate in mass slaughter on the battlefield or in a triumphal march through the enemy's capital. Indeed, he realized that in some circumstances, the absence of a decisive outcome either way was already sufficient for one side: 'the very lack of a decision' might thus constitute 'a success for the defence' (p. 175). As Henry Kissinger remarked a century later, 'The conventional army loses if it does not win. The guerrilla wins if he does not lose.'[43] A superior power may lose a war if it does not manage to impose its will on a population, and a people may manage to persuade a militarily largely superior occupying force to withdraw by denying it the fruits of its occupation, by ensuring a constant haemorrhage caused by pin-pricks and terrorist attacks, and never letting the adversary find peace. To put this in Clausewitzian terms, however many battles Napoleon won, he was defeated in the end, as he had not managed to impose his military occupation on the rest of Europe permanently.

Two of Clausewitz's great theories have already been discussed: his related belief in the superiority of the defensive over the offensive, developed in Book 6; and his focus on war as a function of the policies pursued by the entity fighting it. Policies, the political war aims, however, were not the only variable Clausewitz identified as determining the many manifestations of war.

War as a function of the trinities

In developing further his notion of war as a function of other variables, Clausewitz identified a 'remarkable trinity' of variables, all of which could be more or less pronounced, and could therefore determine the shape of any war. This 'remarkable trinity' presents the culmination of his reflections in Book 1, probably the most important and original part of *On War*, the book he had revised to his satisfaction before he had to leave his unfinished manuscript behind. This trinity he described as being composed

---

[43] Henry Kissinger: 'The Vietnam Negotiations', *Foreign Affairs*, 48/2 (Jan. 1969), 214.

of primordial violence, hatred, and enmity, which are to be regarded as a blind natural force; of the play of chance and probability within which the creative spirit is free to roam; and of its element of subordination, as an instrument of policy, which makes it subject to reason alone. (p. 30)

Of these three dimensions, that of violence-hatred-enmity he associated with the passions of the people as a whole, i.e. the more the people were involved in a war, the more they identified with it as the French had done with the Revolutionary Wars and the Russians and Prussians with their wars against the hateful Napoleon, the more violent the war would be. The second dimension, that of probability and chance, he associated with 'the interplay of courage and talent' that depended on the peculiarities of the military commander and the army. The third dimension, the political purpose of war, he defined as the will of the government alone. His theory was that any war is shaped by the interplay of all three dimensions, that it was a function of all three sets of variables (pp. 30–1).

From this, subsequent generations of thinkers have derived the concept of a trinity of government, military, and population (society) as a fundamental analytical tool for the study of war. Yet others, after the end of the Cold War and the renewed prominence of warfare with non-state actors (for example guerrilla forces, insurgents, terrorist groupings), have taken this derivative trinity of government/military/population as a sign of Clausewitz's outdatedness, since rebel forces, or warlords, could not be described in the neat categories of a government, a (professional) military, and a distinct population. One could also argue that the First and Second World Wars eschewed this neat categorization, as the near-total mobilization of the societies in both wars abolished any meaningful distinction between a war-fighting military and the population, as the latter was fully involved in the war effort.

But putting excessive emphasis on Clausewitz's secondary trinity of government/military/population as distinct elements makes nonsense of Clausewitz's intention in formulating this concept. It is his *primary* trinity that supplies the 'three magnets' between which war is moving like a suspended metal object, constantly following their attraction. And these competing magnetic poles are, to put it another way, war's tendency to escalate to ever greater violence, the more the passions of the people are involved; the political restraints counteracting this tendency or the political aims encouraging it; and the skills and genius of the military leaders, the morale of their forces, and the military contingencies they have imagined and prepared for, which may or may not have prepared them for the real wars they engage in. This trinity can be summed up as one of violence-hatred, chance, and political aims, or to put it in Eulerian terms, war is a function of the variables of violence-hatred, of the luck and the skills of the military, and of the aims of the political leadership. These variables in themselves are interconnected: the tendency towards violence may or may not be curtailed by the political leaders, the military may or may not be influenced by the passion (or disinterest) of the population as a whole, the military's victories or defeats may or may not stir up the passions of the people, and the political leadership may or may not have pursued its interests carefully enough to have prepared the military well for its purpose in war. These ideas continue to be brilliant analytical tools.

How to pursue victory

Much of *On War* is made up of less philosophical, more down-to-earth reflections on war. The wars of the Ancien Régime, according to Guibert and Clausewitz, had lacked decisiveness because they did not seek the destruction of the enemy's army. Accordingly, the results of one summer of campaigning could be overturned in the following year, conflicts could be protracted and warfare indecisive. This seemed so different from Napoleon's campaigns, which until 1812 hardly knew a reversal. Accordingly, much of Clausewitz's writing sought to draw lessons above all from Napoleon's campaigns. A large part of *On War* dealing with these aspects has much in common with the Scharnhorst-style manual when discussing 'flank positions', 'base of operations', and 'terrain', the main subject matter of Books 3 to 7, much of which has been omitted in this edition. But in looking at the operations and technicalities of war, he made a number of astute observations which have remained very useful to its understanding. One is the importance of somehow attacking the adversary's centre of gravity, his *Schwerpunkt*. Originally, in his earlier writings and in the earlier parts of *On War*, as exemplified by the quotation from Book 4 above, what Clausewitz meant by this is the attack of the main forces of the adversary in what should become a decisive battle, in which the enemy's army is beaten devastatingly and bloodily, Napoleonic style. Over time, Clausewitz modified his views as to what the enemy's centre of gravity might consist of: the enemy's main armed forces, or indirectly, the enemy's overall morale, his will to continue the struggle? If it was the latter, then again, a victory on the battlefield might not be enough. Perhaps the enemy nation had to be humiliated into admitting defeat by a seizure of their capital, a military victory parade in its main avenue? Perhaps, if the enemy was not a state with a nation and a capital, but a band of insurgents, one had to seize and publicly execute their leader to break the will of the insurgents (Book 8, Chapter 4).

The domination of the enemy's will as the aim of all warfare opened up a further intellectually fruitful concept, that of the prospect of escalation. In order to break an enemy's will to pursue a war further, one might simply have to lead him to the point of threatening something so terrible, giving him a vision of a future so unbearable, that he would rather give up than run the real risk of experiencing it (*On War*, Book 1, Chapter 1, sections 3 and 4). This threat one would later call the threat of escalation (or alternatively, infinite prolongation) of a conflict, which has become an important part of strategic thinking ever since. Further developments based on this concept led to an important element of twentieth-century Western nuclear strategies, which turned on the threat of escalation to a nuclear level. The concept is of greatest possible importance to all wars, however. Clausewitz described the threat of force and its use as a form of bargain, which always had to be backed by the real ability to implement a threat (just as one has to be able ultimately to service one's debts if one wants to be given credit); this again is a crucially important insight into the functioning of war. Napoleon's celebrated victory at Ulm, in which his opponents had chosen to surrender as they realized they were outnumbered and outflanked, had been taken by some as the perfect victory, war without bloodshed. But Clausewitz was very dismissive of this as an ideal: 'The surrender at Ulm was a unique event', he wrote, 'which would not have happened even to Bonaparte if he had not been willing to shed blood.'[44] The threat had to be backed up by real determination to see it through.

---

[44] *On War*, Book 4, Chapter 11, p. 260 (Princeton edn.)—passage omitted in this edition.

If a threat is all too blatantly incredible, the adversary is likely to call one's bluff; for a fairly incredible threat to have an effect, the threat itself must be so frightening that no one would ever dare call the bluff. That, as French Clausewitzians observed a century later, was how nuclear deterrence worked: it is a product of the threat and its credibility.[45] But once the credibility reaches zero, even the greatest threat no longer affects the enemy's will.

A very important practical insight which Clausewitz gleaned from his extensive experience of campaigns was what he called the 'friction' that stands between any military plan and its realization. Already in looking back at the Russian campaign, he had noted that in actual war everything was much more difficult than it looked in the abstract, and the implementation of plans would be very different from the plans themselves, just as walking, or practising the movements of swimming on land is from trying to walk, or actually swimming, in water (Book 1, Chapter 7). This factor, friction, or chance events, and the confusion of real battle leading to the 'fog of war' (p. 89), are two Clausewitzian concepts particularly dear to military practitioners, as they articulate succinctly the vast chasm between 'war on paper' and war in reality, as Clausewitz put it.

The perfect book on war?

There is general agreement to this day that Clausewitz's *On War* is probably one of the two best books ever written on the subject, together with the classical Chinese writer Sun Tsu's *The Art of War* (for some decades now another favourite with military academies around the world). This by no means makes Clausewitz's work perfect. Quite apart from the incoherence we find in the work on account of the contradictions noted above, it has many shortcomings. There are types of warfare he did not cover in this book, even though he had written about them in his other works, such as partisan warfare (warfare by irregular troops), or problems of logistics (about which he wrote much in his account of the 1812 campaign). He did not write at all about naval warfare, and unlike Scharnhorst or Vauban and others before him, he wasted little time on the technology of his day and age, perhaps fortunately so, as technology is

---

[45] François de Rose, 'NATO and Nuclear Weapons', *Strategic Review*, 15/4 (August), 28, 34.

swiftly overtaken by new developments. Clausewitz in no way fore-
saw the major breakthroughs of the nineteenth and twentieth centuries, from the railway system for the transport of troops and the
telegraph system enabling almost real-time communication across
large distances, to the staggering improvements of the range and
firepower of guns and other explosive weapons, culminating in 1945
with the first test and then use of nuclear weapons. Clausewitz did
not foresee the corresponding growth of battlefields, which by the
First World War could stretch across an entire continent. Accordingly, the sections of *On War* dealing with technical, practical aspects
of warfare, for example how to ford a river or how to march at night,
are now of least interest in the light of modern technologies such as
river-crossing equipment, helicopters, or satellite intelligence, and
have therefore been omitted in this abridged edition.

Clausewitz did not foresee or imagine that ideologies would consider the defeat of armed forces in the field insufficient to define
victory, and would call for the deliberate targeting of the enemy's
civilian populations. Genocide was not part of his mental framework.
The word 'terrorism' does not occur in *On War*. And there have been
many wars since his times, notably the First World War and the
Second World War on the German side, which were not carefully
directed tools of policies: these were wars in which no clear political
aims were defined before embarking upon them, and where political
aims, more or less absurd and detached from the conduct of war
itself, were made up as the conflict evolved, and where the main
driving force behind the war was the bellicose, militaristic cultures of
one or more of the peoples involved.

But few of Clausewitz's insights or theories are truly outdated
by subsequent developments in the history of war, and many of
his insights have proved to be timeless. Of particular value is
the concept of war as a function of several, partly interconnected
variables.[46] The key variables Clausewitz identified—above all the
trinity of violence-hatred, chance, and political aims—and his central definition of war are rocks on which many thinkers since have
consciously or unconsciously built their own concepts of strategy
and international affairs.

[46] As the mathematician Euler would have put it, war (y) is a function (f) of several,
partly interconnected variables ($x_1, x_2, x_3, \ldots x_n$).

NOTE ON THE TEXT

CARL VON CLAUSEWITZ wrote *On War* between 1815 and 1830. He drew on scripts he had produced earlier, but most of his ideas were developed for the first time in this book. In 1827, he had still not completed the last sections of *On War* ('Books' 7 and 8) and also set out to revise the existing scripts; while he completed Books 7 and 8, he never completed the revision of his entire manuscript. He had to put it down in 1830, when he was mobilized to fend off a potential revolt in the Hohenzollerns' Polish territories, where he died in 1831.

It fell to his widow to publish his works posthumously, in ten volumes, between 1832 and 1837. *On War* made up three of these, published 1832–4. Marie von Clausewitz, née von Brühl, failed to add the revisions which Clausewitz had still hoped to make. In consequence, *On War* is an unfinished book, and contains problematic contradictions which have led to many aberrant developments of Clausewitz's theories by later thinkers, and have caused great harm to his reputation.

Ten years later, long excerpts were first published in French, followed by further French translations over the years. An English translation, by Colonel James John Graham, was first produced in 1873, and this was also published in North America. The Second World War provided renewed incentives to read *On War*, and in 1943 a new English translation was produced by Professor O. J. Matthijs Jolles. In the post-war years, a gifted German historian set out to edit the work afresh: Werner Hahlweg discovered several mistakes and interesting changes made in later German editions of *On War* on which some of the earlier translations had been based. The translation used in the present edition is based on the first edition of 1832, supplemented by Hahlweg's annotated German text. It was undertaken by a British and an American scholar respectively, Michael Howard and Peter Paret, initially with the help of a British diplomat, Angus Malcolm, who died before the volume was finished. This translation was first published in 1976, marking the beginning of a great revival of interest in Clausewitz, especially among the US military, after what was widely seen as a neglect of sound strategic principles in the Vietnam War.

This abridgement seeks to include as much as possible of Clausewitz's original and timeless thinking, contained mainly in Books 1, 2, and 7, 8, while omitting large parts of Books 3–4, which follow a classical pattern of officer's manuals on the art of war. These are both much less original and in large part overtaken by developments in technology since Clausewitz's time. The footnotes are Clausewitz's own; asterisks indicate the editor's explanatory notes at the back of the book.

SELECT BIBLIOGRAPHY

German editions of Clausewitz's work

Vom Kriege, ed. Werner Hahlweg (19th edn., Bonn: Ferd. Dümmler Verlag, 1991).

Carl von Clausewitz: Schriften, Aufsätze, Studien, Briefe, ed. Werner Hahlweg, vol. i (Göttingen: Vandenhoeck & Rupprecht, 1966), vol. ii (Göttingen: Vandenhoeck & Rupprecht, 1990).

Clausewitz: Verstreute Kleine Schriften, ed. Werner Hahlweg, vols. i–iii (Osnabrück: Biblio Verlag, 1979).

Carl von Clausewitz: Strategie aus dem Jahr 1804 mit Zusätzen von 1808 und 1809, ed. Eberhard Kessel (2nd edn., Hamburg: Hanseatische Verlagsanstalt, 1943).

Carl von Clausewitz: Politische Schriften und Briefe, ed. Hans Rothfels (Munich: Drei Masken Verlag, 1922).

Karl und Marie von Clausewitz: Ein Lebensbild in Briefen und Tagebuchblättern, ed. Karl Linnebach (Berlin: Martin Warneck, 1917).

English translations

On War, trans. and ed. Michael Howard and Peter Paret (Princeton: Princeton University Press, 1976).

'The Most Important Principles for the Conduct of War' (1810–12), trans. Hans W. Gatzke in *Roots of Strategy*, Book II (Mechanicsburg, PA: Stackpole Books, 1987).

The Campaign of 1812 in Russia, trans. John Murray (London, 1843; repr. London: Greenhill, 1992).

Secondary literature

Aron, Raymond, *Clausewitz, Philosopher of War*, trans. Christine Booker and Norman Stone (London: Routledge & Kegan Paul, 1976).

Bassford, Christopher, *Clausewitz in English: The Reception of Clausewitz in Britain and America 1815–1945* (Oxford: Oxford University Press, 1994).

Brodie, Bernard, 'The Continuing Relevance of *On War*', in *Carl von Clausewitz: On War*, ed. and trans. Michael Howard and Peter Paret (Princeton: Princeton University Press, 1976).

Carlyle, Robert, *Clausewitz' Contemporary Relevance*, Strategic and Combat Studies Occasional Papers No. 16 (1995).

Chickering, Roger, 'Total War: The Use and Abuse of a Concept', in Manfred Boemeke, Roger Chickering, and Stig Förster (eds.),

Anticipating Total War: The German and American Experiences (Cambridge, Mass.: Harvard University Press, 1999), 13–29.

—— and Förster, Stig (eds.), *On the Road to Total War* (Cambridge, Mass.: Harvard University Press, 1993).

Cimbala, Stephen J., *Clausewitz and Chaos: Friction in War and Military Policy* (Westpoint, Conn: Praeger, 2001).

—— *Clausewitz and Escalation: Classical Perspectives on Nuclear Strategy* (London: Cass, 1991).

Delbrück, Hans, *Friedrich der Große und Clausewitz* (Berlin: H. Peters, 1892).

Dill, Günter (ed.), *Clausewitz in Perspektive* (Frankfurt am Main: Ullstein, 1980).

Echevaria II, Antuilo J., 'Understanding the Art of War: Clausewitz and Martial Principles', *Clausewitz-Studien* (1998), 51–7.

Esdaile, Charles J., *The Wars of Napoleon* (London: Longman, 1995).

Gat, Azar, *The Origins of Military Thought* (Oxford: Clarendon Press, 1989).

Gersdorff, Ursula von (ed.) *Geschichte und Militärgeschichte* (Frankfurt am Main: Bernard Graefe, 1974).

Hagemann, Ernst, *Die deutsche Lehre vom Kriege: Von Berenhorst zu Clauseswitz* (Berlin: E. S. Mittler & Sohn, 1940).

Hahlweg, Werner, *Carl von Clausewitz: Soldat, Politiker, Denker* (Göttingen: Musterschmidt Verlag, 1957).

Handel, Michael I. (ed.), *Clausewitz and Modern Strategy* (London, Frank Cass, 1986).

—— *Masters of War: Classical Strategic Thought* (London: Frank Cass, 1992; 2nd edn. 1996).

Herberg-Rothe, Andreas, *Das Rätsel Clausewitz: Politische Theorie des Krieges im Widerstreit* (Munich: Fink, 2001).

—— and Strachan, Hew (eds.), *Clausewitz in the 21st century* (forthcoming, 2006).

Heuser, Beatrice, *Reading Clausewitz* (London: Pimlico, 2002).

Honig, Jan Willem, 'Interpreting Clausewitz', *Security Studies*, 3/3 (Spring 1994), 571–80.

Kessel, Eberhard, 'Zur Entstehungsgeschichte von Clausewitz' Werk vom Kriege', *Historische Zeitschrift*, 151 (1935), 97–100.

—— 'Zur Genesis der modernen Kriegslehre', *Wehrwissenschaftliche Rundschau*, 3/9 (1953), 405–23.

Kitchen, Martin, 'The Political History of Clausewitz', *Journal of Strategic Studies*, 11/1 (1988), 27–50.

Kondylis, Panajotis, *Theorie des Krieges: Clausewitz—Marx—Engels—Lenin* (Stuttgart: Klett Cotta, 1988).

Marwedel, Ulrich, *Carl von Clausewitz: Persönlichkeit und Wirkungs-geschichte seines Werkes bis 1918* (Boppard am Rhein: Harald Boldt Verlag, 1978).

Murray, Stewart L., *The Reality of War: A Companion to Clausewitz* (London: Hodder & Stoughton, 1914).

Murray, Williamson, Knox, Macgregor, and Bernstein, Alvin (eds.), *The Making of Strategy: Rulers, States, and War* (Cambridge: Cambridge University Press, 1994).

Paret, Peter, *Clausewitz and the State* (Oxford: Clarendon Press, 1976).

Parkinson, Roger, *Clausewitz: A Biography* (London: Wayland, 1970).

Reemtsma, Jan Philipp, 'Die Idee des Vernichtungskrieges: Clausewitz— Ludendorff—Hitler', in Hannes Heer and Klaus Naumann (eds.), *Vernichtungskrieg: Verbrechen der Wehrmacht* (Hamburg: HIS, 1995), 377–401.

Ritter, Gerhard, *The Sword and the Sceptre*, trans. Heinz Norden (London: Allen Lane, 1972; 2nd edn. 1973).

Rogers, Clifford J., 'Clausewitz, Genius and the Rules', *Journal of Military History*, 66 (2002), 1167–76.

Romer, Jean-Christophe, 'Quand L'Armée Rouge critiquait Clausewitz', *Stratégique*, 33 (1987), 97–111.

Rose, Olaf, *Clausewitz: Wirkungsgeschichte seines Werkes in Russland und der Sowjetunion 1836–1991* (Munich: Oldenbourg, 1995).

Rosinski, Herbert, 'Die Entwicklung von Clausewitz' Werk *Vom Kriege* im Lichte seiner Vorreden und Nachrichten', *Historische Zeitschrift*, 151 (1935), 278–93.

Rothfels, Hans (ed.), *Carl von Clausewitz: Politik und Krieg. Eine Ideen-geschichtliche Studie* (Berlin: Dümmler, 1920).

Schering, Walter Malmsten (ed.), *Clausewitz: Geist und Tat* (Stuttgart: Alfred Kröner, 1941).

Schwartz, Karl, *Leben des Generals Carl von Clausewitz und der Frau Marie von Clausewitz geborene Gräfin von Brühl mit Briefen, Aufsätzen und anderen Schriftstücken* (Berlin: Dümmler, 1878).

Stamp, Gerd, *Clausewitz im Nuklearzeitalter* (Wiesbaden: Rheinische Verlagsanstalt, 1962).

Tashjean, John E., *Transatlantic Clausewitz* (Carlisle Baracks, PA: Strategic Studies Institute, US Army War College, 1982).

Terray, Emmanuel, *Clausewitz* (Paris: Fayard, 1999).

Villacres, Edward J., and Bassford, Christopher, 'Reclaiming the Clause-witzian Trinity', *Parameters* (Autumn 1995).

Wallach, Jehuda L., *The Dogma of the Battle of Annihilation: The Theories of Clausewitz and Schlieffen and their Effect on the German Conduct of Two World Wars* (Westport, Conn.: Praeger, 1988).

Further Reading in Oxford World's Classics

Livy, *Hannibal's War*, trans. J. C. Yardley, ed. Dexter Hoyos.
Tolstoy, Leo, *War and Peace*, trans. Louise and Aylmer Maude, ed. Henry
  Gifford.

A CHRONOLOGY OF CLAUSEWITZ AND EUROPEAN CONFLICTS

1700–21   Northern War.

1701–13   War of the Spanish Succession.

1712   Birth of Frederick, son of Frederick William I of Prussia, the 'Soldier-King'.

1740   Upon the death of his father Frederick William I, Frederick II becomes king of Prussia. The death of Emperor Charles VI gives rise to the War of the Austrian Succession (1740–8), when his daughter Maria Theresa's succession is challenged by the elector of Bavaria, who becomes Charles VII (1742–5).

1740–2   First Silesian War between Prussia and Austria.

1744–5   Second Silesian War between Prussia and Austria.

1745   The husband of Maria Theresa, Francis III Stephen of Lorraine, becomes Emperor Francis I (dies 1765).

1756–63   Seven Years War.

1765–90   Maria Theresa's son rules as Emperor Joseph II.

1769   Birth of Napoleon Bonaparte.

1780   1 June: Carl Philipp Gottlieb von Clausewitz is born in Burg near Magdeburg in Brandenburg, which belongs to the Hohenzollern kings of Prussia. 29 November: death of Maria Theresa.

1786   Death of Frederick II (the Great) of Prussia; succession of his nephew, Frederick William II.

1789   Beginning of the French Revolution.

1790–2   Maria Theresa's second son rules as Emperor Leopold II.

1791   Jacques Antoine Count Guibert dies.

1792   Beginning of the French Revolutionary Wars with the War of the First Coalition (1792–7).

1792   Beginning of French Revolutionary Wars: first Coalition War of Austria and Prussia against France 1792–7. Clausewitz joins Prussian army as an officer cadet. 1 March: Francis II succeeds his father as emperor of the Holy Roman Empire. 20 September: French victory at Valmy.

1793   *Levée en masse*: massive French recruitment of volunteers in a Republican spirit.

1793, 1794 Clausewitz takes part in wars with France; witnesses burning of Mainz.

1795–1801 Stationed in Neuruppin.

1799–1802 War of the Second Coalition against France.

1802–4 Clausewitz studies in Scharnhorst's Academy for Young Officers.

1803 At court, he meets Marie, Countess von Brühl, who successively holds several offices in the royal households.

1804 11 August: Francis II assumes the title of Francis I, emperor of Austria (dies 1835). 2 December: Napoleon crowns himself emperor.

1805 Frederick William II of Prussia dies; is succeeded by his son Frederick William III (dies 1840).

1805 War of the Third Coalition. 20 October: Austrian forces surrender to Napoleon at Ulm; 12 November: Napoleon seizes Vienna. 2 December: Napoleon defeats Austrian and Russian forces at Austerlitz.

1806 Fourth Coalition War, leading to the end of the Holy Roman Empire. 14 October: Clausewitz witnesses battle of Jena and Auerstedt, where Napoleon defeats Prussia. 28 October: witnesses the following encounter at Prenzlau. Taken prisoner together with Prince Augustus Ferdinand, and is taken to France.

1807 Peace of Tilsit between Prussia and France. Augustus Ferdinand and Clausewitz are released and travel back to Berlin. Clausewitz joins Scharnhorst in Königsberg (now Kaliningrad).

1808–14 Napoleon's war on the Spanish Peninsula (Peninsular War); Spanish *guerrilla*.

1809 Napoleon goes to war against Austria; 1809–10 Tyrolean popular insurrection.

1810 Clausewitz marries Marie, Countess von Brühl. No children.

1810–12 Clausewitz teaches at the General Military School in Berlin, is tutor to the crown prince Frederic William (later IV).

1812 February: Franco-Prussian pact against Russia. Outraged, Clausewitz writes plans for popular insurrection, resigns his commission, and goes into Russian service. He witnesses Napoleon's campaign of 1812 in Russia, especially the battles of Borodino and Smolensk.

1813–15   Takes part in the 'wars of liberation' against Napoleon.

1813   16–18 October: 'Battle of the Nations' at Leipzig; Napoleon is defeated.

1814   31 March: Coalition marches into Paris. 6 April: Napoleon abdicates and is exiled on Elba.

1814–15   Vienna Congress inaugurates period of the European Concert.

1815   1 March: Napoleon returns to France. 18 June: Battle of Belle Alliance (Waterloo) ends in Napoleon's defeat. Clausewitz is present near the main battlefield.

1815–30   Clausewitz becomes a general and administrative director of General War School. During this time he writes *On War*.

1820s   Return of limited wars, with great-power interventions in Italy, Spain, and Russian clashes with Turkey (Russian-Turkish War 1828–9).

1821   Death of Napoleon.

1830   Spring: Clausewitz leaves the General War School and joins the artillery. June: Revolution in France. Polish insurrection against Russian rule.

1831   March: Clausewitz is deployed to Prussian-held Polish territories to prevent a Polish insurrection; 16 November: dies of cholera in Breslau/Wrocław.

1832–4   Clausewitz's widow publishes *On War* in three volumes, while resuming posts in the royal households as *Oberhofmeisterin* (principal lady-in-waiting) and as governess to the future Emperor Frederic I.

ON WAR

CONTENTS

TWO NOTES BY THE AUTHOR ON HIS PLANS
FOR REVISING *ON WAR*

Note of 10 July 1827

I regard the first six books, which are already in a clean copy, merely as a rather formless mass that must be thoroughly reworked once more. The revision will bring out the two types of war with greater clarity at every point. All ideas will then become plainer, their general trend will be more clearly marked, their application shown in greater detail.

War can be of two kinds, in the sense that either the objective is to *overthrow the enemy*—to render him politically helpless or militarily impotent, thus forcing him to sign whatever peace we please; or *merely to occupy some of his frontier-districts* so that we can annex them or use them for bargaining at the peace negotiations. Transitions from one type to the other will of course recur in my treatment; but the fact that the aims of the two types are quite different must be clear at all times, and their points of irreconcilability brought out.

This distinction between the two kinds of war is a matter of actual fact. But no less practical is the importance of another point that must be made absolutely clear, namely that *war is nothing but the continuation of policy with other means*. If this is firmly kept in mind throughout it will greatly facilitate the study of the subject and the whole will be easier to analyse. Although the main application of this point will not be made until Book Eight, it must be developed in Book One and will play its part in the revision of the first six books. That revision will also rid the first six books of a good deal of superfluous material, fill in various gaps, large and small, and make a number of generalities more precise in thought and form.

Book Seven, 'On Attack' (various chapters of which are already in rough draft) should be regarded as the counterpart of Book Six, 'On Defence', and is the next to be revised in accordance with the clear insights indicated above. Thereafter it will need no further revision; indeed, it will then provide a standard for revising the first six books.

Book Eight, 'War-Plans', will deal with the organization of a war as a whole. Several chapters of it have already been drafted, but they must not in any sense be taken as being in final form. They are really no more than a rough working over of the raw material, done with the idea that the labour itself would show what the real problems were. That in fact is what happened, and when I have finished Book Seven I shall go on at once and work out Book Eight in full. My main concern will be to apply the two principles mentioned above, with the idea of refining and simplifying everything. In Book Eight I also hope to iron out a good many kinks in the minds of strategists and statesmen and at all events to show what the whole thing is about and what the real problems are that have to be taken into account in actual warfare.

If the working out of Book Eight results in clearing my own mind and in really establishing the main features of war it will be all the easier for me to apply the same criteria to the first six books and make those features evident throughout them. Only when I have reached that point, therefore, shall I take the revision of the first six books in hand.

If an early death should terminate my work, what I have written so far would, of course only deserve to be called a shapeless mass of ideas. Being liable to endless misinterpretation it would be the target of much half-baked criticism, for in matters of this kind everyone feels he is justified in writing and publishing the first thing that comes into his head when he picks up a pen, and thinks his own ideas as axiomatic as the fact that two and two make four. If critics would go to the trouble of thinking about the subject for years on end and testing each conclusion against the actual history of war, as I have done, they would undoubtedly be more careful of what they said.

Nonetheless, I believe an unprejudiced reader in search of truth and understanding will recognize the fact that the first six books, for all their imperfection of form, contain the fruit of years of reflection on war and diligent study of it. He may even find they contain the basic ideas that might bring about a revolution in the theory of war.

Unfinished Note, Presumably Written in 1830

The manuscript on the conduct of major operations that will be found after my death can, in its present state, be regarded as nothing but a collection of materials from which a theory of war was to have been distilled. I am still dissatisfied with most of it, and can call Book

Six only a sketch. I intended to rewrite it entirely and to try and find a solution along other lines.

Nevertheless I believe the main ideas which will be seen to govern this material are the right ones, looked at in the light of actual warfare. They are the outcome of wide-ranging study: I have thoroughly checked them against real life and have constantly kept in mind the lessons derived from my experience and from association with distinguished soldiers.

Book Seven, which I have sketched in outline, was meant to deal with 'Attack', and Book Eight with 'War-Plans', in which I intended to concern myself particularly with war in its political and human aspects.

The first chapter of Book One alone I regard as finished. It will at least serve the whole by indicating the direction I meant to follow everywhere.

The theory of major operations (strategy, as it is called) presents extraordinary difficulties, and it is fair to say that very few people have clear ideas about its details—that is, ideas which logically derive from basic necessities. Most men merely act on instinct, and the amount of success they achieve depends on the amount of talent they were born with.

All great commanders have acted on instinct, and the fact that their instinct was always sound is partly the measure of their innate greatness and genius. So far as action is concerned this will always be the case and nothing more is needed. Yet when it is not a question of acting oneself but of persuading others in discussion, the need is for clear ideas and the ability to show their connection with each other. So few people have yet acquired the necessary skill at this that most discussions are a futile bandying of words; either they leave each man sticking to his own ideas or they end with everyone agreeing, for the sake of agreement, on a compromise with nothing to be said for it.

Clear ideas on these matters do, therefore, have some practical value. The human mind, moreover, has a universal thirst for clarity, and longs to feel itself part of an orderly scheme of things.

It is a very difficult task to construct a scientific theory for the art of war, and so many attempts have failed that most people say it is impossible, since it deals with matters that no permanent law can provide for. One would agree, and abandon the attempt, were it not for the obvious fact that a whole range of propositions can be demonstrated without difficulty: that defence is the stronger form of fighting with the negative purpose, attack the weaker form with the positive purpose; that major successes help bring about minor ones, so that strategic results can be traced back to certain turning-points; that a demonstration is a weaker use of force than a real attack, and that it must therefore be clearly justified; that victory consists not only in the occupation of the battlefield, but in the destruction of the enemy's physical and psychic forces, which is usually not attained until the enemy is pursued after a victorious battle; that success is always greatest at the point where the victory was gained, and that consequently changing from one line of operations, one direction, to another can at best be regarded as a necessary evil; that a turning movement can only be justified by general superiority or by having better lines of communication or retreat than the enemy's; that flank-positions are governed by the same consideration; that every attack loses impetus as it progresses.

BOOK ONE
ON THE NATURE OF WAR

CHAPTER I

WHAT IS WAR?

1. Introduction

I PROPOSE to consider first the various *elements* of the subject, next its *various parts* or *sections*, and finally *the whole* in its internal structure. In other words, I shall proceed from the simple to the complex. But in war more than in any other subject we must begin by looking at the nature of the whole; for here more than elsewhere the part and the whole must always be thought of together.

2. Definition

I shall not begin by expounding a pedantic, literary definition of war, but go straight to the heart of the matter, to the duel. War is nothing but a duel on a larger scale. Countless duels go to make up war, but a picture of it as a whole can be formed by imagining a pair of wrestlers. Each tries through physical force to compel the other to do his will; his *immediate* aim is to *throw* his opponent in order to make him incapable of further resistance.

War is thus an act of force to compel our enemy to do our will.

Force, to counter opposing force, equips itself with the inventions of art and science. Attached to force are certain self-imposed, imperceptible limitations hardly worth mentioning, known as international law and custom, but they scarcely weaken it. Force—that is, physical force, for moral force has no existence save as expressed in the state and the law—is thus the *means* of war; to impose our will on the enemy is its *object*. To secure that object we must render the enemy powerless; and that, in theory, is the true aim of warfare. That aim takes the place of the object, discarding it as something not actually part of war itself.

3. The Maximum Use of Force

Kind-hearted people might of course think there was some ingenious way to disarm or defeat an enemy without too much bloodshed, and might imagine this is the true goal of the art of war. Pleasant as it sounds, it is a fallacy that must be exposed: war is such a dangerous business that the mistakes which come from kindness are the very worst. The maximum use of force is in no way incompatible with the simultaneous use of the intellect. If one side uses force without compunction, undeterred by the bloodshed it involves, while the other side refrains, the first will gain the upper hand. That side will force the other to follow suit; each will drive its opponent toward extremes, and the only limiting factors are the counterpoises inherent in war.

This is how the matter must be seen. It would be futile—even wrong—to try and shut one's eyes to what war really is from sheer distress at its brutality.

If wars between civilized nations are far less cruel and destructive than wars between savages, the reason lies in the social conditions of the states themselves and in their relationships to one another. These are the forces that give rise to war; the same forces circumscribe and moderate it. They themselves however are not part of war; they already exist before fighting starts. To introduce the principle of moderation into the theory of war itself would always lead to logical absurdity.

Two different motives make men fight one another: *hostile feelings* and *hostile intentions*. Our definition is based on the latter, since it is the universal element. Even the most savage, almost instinctive, passion of hatred cannot be conceived as existing without hostile intent; but hostile intentions are often unaccompanied by any sort of hostile feelings—at least by none that predominate. Savage peoples are ruled by passion, civilized peoples by the mind. The difference, however, lies not in the respective natures of savagery and civilization, but in their attendant circumstances, institutions, and so forth. The difference, therefore, does not operate in every case, but it does in most of them. Even the most civilized of peoples, in short, can be fired with passionate hatred for each other.

Consequently, it would be an obvious fallacy to imagine war between civilized peoples as resulting merely from a rational act on the part of their governments and to conceive of war as gradually ridding itself of passion, so that in the end one would never really need to use the physical impact of the fighting forces—comparative figures of their strength would be enough. That would be a kind of war by algebra.

Theorists were already beginning to think along such lines when the recent wars* taught them a lesson. If war is an act of force, the emotions cannot fail to be involved. War may not spring from them, but they will still affect it to some degree, and the extent to which they do so will depend not on the level of civilization but on how important the conflicting interests are and on how long their conflict lasts.

If, then, civilized nations do not put their prisoners to death or devastate cities and countries, it is because intelligence plays a larger part in their methods of warfare and has taught them more effective ways of using force than the crude expression of instinct.

The invention of gunpowder and the constant improvement of firearms are enough in themselves to show that the advance of civilization has done nothing practical to alter or deflect the impulse to destroy the enemy, which is central to the very idea of war.

The thesis, then, must be repeated: war is an act of force, and there is no logical limit to the application of that force. Each side, therefore, compels its opponent to follow suit; a reciprocal action is started which must lead, in theory, to extremes. This is the *first case of interaction and the first 'extreme'* we meet with.

4. *The Aim Is To Disarm the Enemy*

I have already said that the aim of warfare is to disarm the enemy and it is time to show that, at least in theory, this is bound to be so. If the enemy is to be coerced you must put him in a situation that is even more unpleasant* than the sacrifice you call on him to make. The hardships of that situation must not of course be merely transient— at least not in appearance. Otherwise the enemy would not give in but would wait for things to improve. Any change that might be brought about by continuing hostilities must then, at least in theory, be of a kind to bring the enemy still greater disadvantages. The worst of all conditions in which a belligerent can find himself is to be utterly defenceless. Consequently, if you are to force the enemy, by making war on him, to do your bidding, you must either make him literally defenceless or at least put him in a position that makes this danger probable. It follows, then, that to overcome the enemy, or disarm him—call it what you will—must always be the aim of warfare.

War, however, is not the action of a living force upon a lifeless mass (total nonresistance would be no war at all) but always the collision of two living forces. The ultimate aim of waging war, as formulated here, must be taken as applying to both sides. Once again, there is interaction. So long as I have not overthrown my opponent I am bound to fear he may overthrow me. Thus I am not in control: he dictates to me as much as I dictate to him. This is the *second case of interaction and it leads to the second 'extreme'*.

5. The Maximum Exertion of Strength

If you want to overcome your enemy you must match your effort against his power of resistance, which can be expressed as the product of two inseparable factors, viz. *the total means at his disposal* and *the strength of his will*. The extent of the means at his disposal is a matter—though not exclusively—of figures, and should be measurable. But the strength of his will is much less easy to determine and can only be gauged approximately by the strength of the motive animating it. Assuming you arrive in this way at a reasonably accurate estimate of the enemy's power of resistance, you can adjust your own efforts accordingly; that is, you can either increase them until they surpass the enemy's or, if this is beyond your means, you can make your efforts as great as possible. But the enemy will do the same; competition will again result and, in pure theory, it must again force you both to extremes. This is *the third case of interaction and the third 'extreme'*.

6. Modifications in Practice

Thus in the field of abstract thought the inquiring mind can never rest until it reaches the extreme, for here it is dealing with an extreme: a clash of forces freely operating and obedient to no law but their own. From a pure concept of war you might try to deduce absolute terms for the objective you should aim at and for the means of achieving it; but if you did so the continuous interaction would land you in extremes that represented nothing but a play of the imagination issuing from an almost invisible sequence of logical subtleties. If we were to think purely in absolute terms, we could avoid every difficulty by a stroke of the pen and proclaim with inflexible logic that, since the extreme must always be the goal, the greatest effort must always be exerted. Any such pronouncement would be an abstraction and would leave the real world quite unaffected.

Even assuming this extreme effort to be an absolute quantity that could easily be calculated, one must admit that the human mind is unlikely to consent to being ruled by such a logical fantasy. It would often result in strength being wasted, which is contrary to other principles of statecraft. An effort of will out of all proportion to the object in view would be needed but would not in fact be realized, since subtleties of logic do not motivate the human will.

But move from the abstract to the real world, and the whole thing looks quite different. In the abstract world, optimism was all-powerful and forced us to assume that both parties to the conflict not only sought perfection but attained it. Would this ever be the case in practice? Yes, it would if: (a) war were a wholly isolated act, occurring suddenly and not produced by previous events in the political world; (b) it consisted of a single decisive act or a set of simultaneous ones; (c) the decision achieved was complete and perfect in itself, uninfluenced by any previous estimate of the political situation it would bring about.

7. War Is Never an Isolated Act

As to the first of these conditions, it must be remembered that neither opponent is an abstract person to the other, not even to the extent of that factor in the power of resistance, namely the will, which is dependent on externals. The will is not a wholly unknown factor; we can base a forecast of its state tomorrow on what it is today. War never breaks out wholly unexpectedly, nor can it be spread instantaneously. Each side can therefore gauge the other to a large extent by what he is and does, instead of judging him by what he, strictly speaking, ought to be or do. Man and his affairs, however, are always something short of perfect and will never quite achieve the absolute best. Such shortcomings affect both sides alike and therefore constitute a moderating force.

8. War Does Not Consist of a Single Short Blow

The second condition calls for the following remarks:

If war consisted of one decisive act, or of a set of simultaneous decisions, preparations would tend toward totality, for no omission could ever be rectified. The sole criterion for preparations which the world of reality could provide would be the measures taken by the adversary—so far as they are known; the rest would once more be reduced to abstract calculations. But if the decision in war consists of several successive acts, then each of them, seen in context, will provide a gauge for those that follow. Here again, the abstract world is ousted by the real one and the trend to the extreme is thereby moderated.

But, of course, if all the means available were, or could be, simultaneously employed, all wars would automatically be confined to a single decisive act or a set of simultaneous ones—the reason being that any *adverse* decision must reduce the sum of the means available, and if *all* had been committed in the first act there could really be no question of a second. Any subsequent military operation would virtually be part of the first—in other words, merely an extension of it.

Yet, as I showed above, as soon as preparations for a war begin, the world of reality takes over from the world of abstract thought; material calculations take the place of hypothetical extremes and, if for no other reason, the interaction of the two sides tends to fall short of maximum effort. Their full resources will therefore not be mobilized immediately.

Besides, the very nature of those resources and of their employment means they cannot all be deployed at the same moment. The resources in question are *the fighting forces proper, the country*, with its physical features and population, and its *allies*.

The country—its physical features and population—is more than just the source of all armed forces proper; it is in itself an integral element among the factors at work in war—though only that part which is the actual theatre of operations or has a notable influence on it.

It is possible, no doubt, to use all mobile fighting forces simultaneously; but with fortresses, rivers, mountains, inhabitants, and so forth, that cannot be done; not, in short, with the country as a whole, unless it is so small that the opening action of the war completely engulfs it. Furthermore, allies do not cooperate at the mere desire of those who are actively engaged in fighting; international relations being what they are, such cooperation is often furnished only at some later stage or increased only when a balance has been disturbed and needs correction.

In many cases, the proportion of the means of resistance that cannot immediately be brought to bear is much higher than might at first be thought. Even when great strength has been expended on the first decision and the balance has been badly upset, equilibrium can be restored. The point will be more fully treated in due course. At this stage it is enough to show that the very nature of war impedes the *simultaneous concentration of all forces*. To be sure, that fact in itself cannot be grounds for making any but a maximum effort to obtain the first decision, for a defeat is always a disadvantage no one would deliberately risk. And even if the first clash is not the only one, the influence it has on subsequent actions will be on a scale proportionate to its own. But it is contrary to human nature to make an extreme effort, and the tendency therefore is always to plead that a decision may be possible later on. As a result, for the first decision, effort and concentration of forces are not all they might be. Anything omitted out of weakness by one side becomes a real, *objective* reason for the other to reduce its efforts, and the tendency toward extremes is once again reduced by this interaction.

9. In War the Result Is Never Final

Lastly, even the ultimate outcome of a war is not always to be regarded as final. The defeated state often considers the outcome merely as a transitory evil, for which a remedy may still be found in political conditions at some later date. It is obvious how this, too, can slacken tension and reduce the vigour of the effort.

10. The Probabilities of Real Life Replace the Extreme and the Absolute Required by Theory

Warfare thus eludes the strict theoretical requirement that extremes of force be applied. Once the extreme is no longer feared or aimed at, it becomes a matter of judgement what degree of effort should be made; and this can only be based on the phenomena of the real world and the *laws of probability*. Once the antagonists have ceased to be mere figments of a theory and become actual states and governments, when war is no longer a theoretical affair but a series of actions obeying its own peculiar laws, reality supplies the data from which we can deduce the unknown that lies ahead.

From the enemy's character, from his institutions, the state of his affairs and his general situation, each side, using the *laws of probability*, forms an estimate of its opponent's likely course and acts accordingly.

11. The Political Object Now Comes to the Fore Again

A subject which we last considered in Section 2 now forces itself on us again, namely the *political object of the war*. Hitherto it had been rather overshadowed by the law of extremes, the will to overcome the enemy and make him powerless. But as this law begins to lose its force and as this determination wanes, the political aim will reassert itself. If it is all a calculation of probabilities based on given individuals and conditions, the *political object*, which was the *original motive*, must become an essential factor in the equation. The smaller the penalty you demand from your opponent, the less you can expect him to try and deny it to you; the smaller the effort he makes, the less you need make yourself. Moreover, the more modest your own political aim, the less importance you attach to it and the less reluctantly you will abandon it if you must. *This is another reason why your effort will be modified.*

The political object—the original motive for the war—will thus determine both the military objective to be reached and the amount of effort it requires. The political object cannot, however, in *itself* provide the standard of measurement. Since we are dealing with realities, not with abstractions, it can do so only in the context of the two states at war. The same political object can elicit *differing* reactions from different peoples, and even from the same people at different times. We can therefore take the political object as a standard only if we think of *the influence it can exert upon the forces it is meant to move*. The nature of those forces therefore calls for study. Depending on whether their characteristics increase or diminish the drive toward a particular action, the outcome will vary. Between two peoples and two states there can be such tensions, such a mass of inflammable material, that the slightest quarrel can produce a wholly disproportionate effect—a real explosion.

This is equally true of the efforts a political object is expected to arouse in either state, and of the military objectives which their policies require. Sometimes the *political and military objective is the same*—for example, the conquest of a province. In other cases the political object will not provide a suitable military objective. In that event, another military objective must be adopted that will serve the political purpose and symbolize it in the peace negotiations. But here, too, attention must be paid to the character of each state involved. There are times when, if the political object is to be achieved, the substitute must be a good deal more important. The less involved the population and the less serious the strains within states and between them, the more political requirements in themselves will dominate and tend to be decisive. Situations can thus exist in which the political object will almost be the sole determinant.

Generally speaking, a military objective that matches the political object in scale will, if the latter is reduced, be reduced in proportion; this will be all the more so as the political object increases its predominance. Thus it follows that without any inconsistency wars can have all degrees of importance and intensity, ranging from a war of extermination down to simple armed observation. This brings us to a different question, which now needs to be analysed and answered.

12. An Interruption of Military Activity Is Not Explained by Anything Yet Said

However modest the political demands may be on either side, however small the means employed, however limited the military objective, can the process of war ever be interrupted, even for a moment? The question reaches deep into the heart of the matter.

Every action needs a certain time to be completed. That period is called its duration, and its length will depend on the speed with which the person acting works. We need not concern ourselves with the difference here. Everyone performs a task in his own way; a slow man, however, does not do it more slowly because he wants to spend more time over it, but because his nature causes him to need more time. If he made more haste he would do the job less well. His speed, then, is determined by subjective causes and is a factor in the actual duration of the task.

Now if every action in war is allowed its appropriate duration, we would agree that, at least at first sight, any additional expenditure of time—any suspension of military action—seems absurd. In this connection it must be remembered that what we are talking about is not the progress made by one side or the other but the progress of military interaction as a whole.

13. Only One Consideration Can Suspend Military Action, and It Seems That It Can Never Be Present on More Than One Side

If two parties have prepared for war, some motive of hostility must have brought them to that point. Moreover so long as they remain under arms (do not negotiate a settlement) that motive of hostility must still be active. Only one consideration can restrain it: *a desire to wait for a better moment before acting.* At first sight one would think this desire could never operate on more than one side since its opposite must automatically be working on the other. If action would bring an advantage to one side, the other's interest must be to wait.

But an absolute balance of forces cannot bring about a standstill, for if such a balance should exist the initiative would necessarily belong to the side with the positive purpose—the attacker.

One could, however, conceive of a state of balance in which the side with the positive aim (the side with the stronger grounds for action) was the one that had the weaker forces. The balance would then result from the combined effects of aim and strength. Were that the case, one would have to say that unless some shift in the balance were in prospect the two sides should make peace. If, however, some alteration were to be foreseen, only one side could expect to gain by it—a fact which ought to stimulate the other into action. Inaction clearly cannot be explained by the concept of balance. The only explanation is that both are waiting for a better time to act. Let us suppose, therefore, that one of the two states has a positive aim—say, the conquest of a part of the other's territory, to use for bargaining at the peace table. Once the prize is in its hands, the political object has been achieved; there is no need to do more, and it can let matters rest. If the other state is ready to accept the situation, it should sue for peace. If not, it must do something; and if it thinks it will be better organized for action in four weeks' time it clearly has an adequate reason for not taking action at once.

But from that moment on, logic would seem to call for action by the other side—the object being to deny the enemy the time he needs for getting ready. Throughout all this I have assumed, of course, that both sides understand the situation perfectly.

14. *Continuity Would Thus Be Brought About in Military Action and Would Again Intensify Everything*

If this continuity were really to exist in the campaign its effect would again be to drive everything to extremes. Not only would such ceaseless activity arouse men's feelings and inject them with more passion and elemental strength, but events would follow more closely on each other and be governed by a stricter causal chain. Each individual action would be more important, and consequently more dangerous.

But war, of course, seldom if ever shows such continuity. In numerous conflicts only a very small part of the time is occupied by action, while the rest is spent in inactivity. This cannot always be an anomaly. Suspension of action in war must be possible; in other words, it is not a contradiction in terms. Let me demonstrate this point, and explain the reasons for it.

15. *Here a Principle of Polarity Is Proposed*

By thinking that the interests of the two commanders are opposed in equal measure to each other, we have assumed a genuine *polarity*. A whole chapter will be devoted to the subject further on, but the following must be said about it here.

The principle of polarity is valid only in relation to one and the same object, in which positive and negative interests exactly cancel one another out. In a battle each side aims at victory; that is a case of true polarity, since the victory of one side excludes the victory of the other. When, however, we are dealing with two different things that have a common relation external to themselves, the polarity lies not in the *things* but in their relationship.

16. Attack and Defence Being Things Different in Kind and Unequal in Strength, Polarity Cannot Be Applied to Them

If war assumed only a single form, namely, attacking the enemy, and defence were nonexistent; or, to put it in another way, if the only differences between attack and defence lay in the fact that attack has a positive aim whereas defence has not, and the forms of fighting were identical; then every advantage gained by one side would be a precisely equal disadvantage to the other—true polarity would exist.

But there are two distinct forms of action in war: attack and defence. As will be shown in detail later, the two are very different and unequal in strength. Polarity, then, does not lie in attack or defence, but in the object both seek to achieve: the decision. If one commander wants to postpone the decision, the other must want to hasten it, always assuming that both are engaged in the same kind of fighting. If it is in A's interest not to attack B now but to attack him in four weeks, then it is in B's interest not to be attacked in four weeks' time, but now. This is an immediate and direct conflict of interest; but it does not follow from this that it would also be to B's advantage to make an immediate attack on A. That would obviously be quite another matter.

17. The Superiority of Defence over Attack Often Destroys the Effect of Polarity, and This Explains the Suspension of Military Action

As we shall show, defence is a stronger form of fighting than attack. Consequently we must ask whether the advantage of *postponing a decision* is as great for one side as the advantage of *defence* is for the other. Whenever it is not, it cannot balance the advantage of defence and in this way influence the progress of the war. It is clear, then, that the impulse created by the polarity of interests may be exhausted in the difference between the strength of attack and defence, and may thus become inoperative.

Consequently, if the side favoured by present conditions is not sufficiently strong to do without the added advantages of the defence, it will have to accept the prospect of acting under unfavourable conditions in the future. To fight a defensive battle under these less favourable conditions may still be better than to attack immediately or to make peace. I am convinced that the superiority of the defensive (if rightly understood) is very great, far greater than appears at first sight. It is this which explains without any inconsistency most periods of inaction that occur in war. The weaker the motives for action, the more will they be overlaid and neutralized by this disparity between attack and defence, and the more frequently will action be suspended—as indeed experience shows.

18. A Second Cause Is Imperfect Knowledge of the Situation

There is still another factor that can bring military action to a standstill: imperfect knowledge of the situation. The only situation a commander can know fully is his own; his opponent's he can know only from unreliable intelligence. His evaluation, therefore, may be mistaken and can lead him to suppose that the initiative lies with the enemy when in fact it remains with him. Of course such faulty appreciation is as likely to lead to ill-timed action as to ill-timed inaction, and is no more conducive to slowing down operations than it is to speeding them up. Nevertheless, it must rank among the natural causes which, *without entailing inconsistency, can bring military activity to a halt.* Men are always more inclined to pitch their estimate of the enemy's strength too high than too low, such is human nature. Bearing this in mind, one must admit that partial ignorance of the situation is, generally speaking, a major factor in delaying the progress of military action and in moderating the principle that underlies it.

The possibility of inaction has a further moderating effect on the progress of the war by diluting it, so to speak, in time by delaying danger, and by increasing the means of restoring a balance between the two sides. The greater the tensions that have led to war, and the greater the consequent war effort, the shorter these periods of inaction. Inversely, the weaker the motive for conflict, the longer the intervals between actions. For the stronger motive increases willpower, and willpower, as we know, is always both an element in and the product of strength.

19. Frequent Periods of Inaction Remove War Still Further from the Realm of the Absolute and Make It Even More a Matter of Assessing Probabilities

The slower the progress and the more frequent the interruptions of military action the easier it is to retrieve a mistake, the bolder will be the general's assessments, and the more likely he will be to avoid theoretical extremes and to base his plans on probability and inference. Any given situation requires that probabilities be calculated in the light of circumstances, and the amount of time available for such calculation will depend on the pace with which operations are taking place.

20. Therefore Only the Element of Chance Is Needed to Make War a Gamble, And That Element Is Never Absent

It is now quite clear how greatly the objective nature of war makes it a matter of assessing probabilities. Only one more element is needed to make war a gamble—chance: the very last thing that war lacks. No other human activity is so continuously or universally bound up with chance. And through the element of chance, guesswork and luck come to play a great part in war.

21. Not Only Its Objective But Also Its Subjective Nature Makes War a Gamble

If we now consider briefly the *subjective nature* of war—the means by which war has to be fought—it will look more than ever like a gamble. The element in which war exists is danger. The highest of all moral qualities in time of danger is certainly *courage*. Now courage is perfectly compatible with prudent calculation but the two differ nonetheless, and pertain to different psychological forces. Daring, on the other hand, boldness, rashness, trusting in luck are only variants of courage, and all these traits of character seek their proper element—chance.

In short, absolute, so-called mathematical, factors never find a firm basis in military calculations. From the very start there is an interplay of possibilities, probabilities, good luck and bad that weaves its way throughout the length and breadth of the tapestry. In the whole range of human activities, war most closely resembles a game of cards.

22. *How in General This Best Suits Human Nature*

Although our intellect always longs for clarity and certainty, our nature often finds uncertainty fascinating. It prefers to day-dream in the realms of chance and luck rather than accompany the intellect on its narrow and tortuous path of philosophical enquiry and logical deduction only to arrive—hardly knowing how—in unfamiliar surroundings where all the usual landmarks seem to have disappeared. Unconfined by narrow necessity, it can revel in a wealth of possibilities; which inspire courage to take wing and dive into the element of daring and danger like a fearless swimmer into the current.

Should theory leave us here, and cheerfully go on elaborating absolute conclusions and prescriptions? Then it would be no use at all in real life. No, it must also take the human factor into account, and find room for courage, boldness, even foolhardiness. The art of war deals with living and with moral forces. Consequently, it cannot attain the absolute, or certainty; it must always leave a margin for uncertainty, in the greatest things as much as in the smallest. With uncertainty in one scale, courage and self-confidence must be thrown into the other to correct the balance. The greater they are, the greater the margin that can be left for accidents. Thus courage and self-confidence are essential in war, and theory should propose only rules that give ample scope to these finest and least dispensable of military virtues, in all their degrees and variations. Even in daring there can be method and caution; but here they are measured by a different standard.

23. *But War Is Nonetheless a Serious Means to a Serious End:*
A More Precise Definition of War

Such is war, such is the commander who directs it, and such the theory that governs it. War is no pastime; it is no mere joy in daring and winning, no place for irresponsible enthusiasts. It is a serious means to a serious end, and all its colourful resemblance to a game of chance, all the vicissitudes of passion, courage, imagination, and enthusiasm it includes are merely its special characteristics.

When whole communities go to war—whole peoples, and especially *civilized* peoples—the reason always lies in some political situation, and the occasion is always due to some political object. War, therefore, is an act of policy. Were it a complete, untrammelled, absolute manifestation of violence (as the pure concept would require), war would of its own independent will usurp the place of policy the moment policy had brought it into being; it would then drive policy out of office and rule by the laws of its own nature, very much like a mine that can explode only in the manner or direction predetermined by the setting. This, in fact, is the view that has been taken of the matter whenever some discord between policy and the conduct of war has stimulated theoretical distinctions of this kind. But in reality things are different, and this view is thoroughly mistaken. In reality war, as has been shown, is not like that. Its violence is not of the kind that explodes in a single discharge, but is the effect of forces that do not always develop in exactly the same manner or to the same degree. At times they will expand sufficiently to overcome the resistance of inertia or friction; at others they are too weak to have any effect. War is a pulsation of violence, variable in strength and therefore variable in the speed with which it explodes and discharges its energy. War moves on its goal with varying speeds; but it always lasts long enough for influence to be exerted on the goal and for its own course to be changed in one way or another—long enough, in other words, to remain subject to the action of a superior intelligence. If we keep in mind that war springs from some political purpose, it is natural that the prime cause of its existence will remain the supreme consideration in conducting it. That, however, does not imply that the political aim is a tyrant. It must adapt itself to its chosen means, a process which can radically change it; yet the political aim remains the first consideration. Policy, then, will permeate all military operations, and, in so far as their violent nature will admit, it will have a continuous influence on them.

24. *War Is Merely the Continuation of Policy by Other Means*

We see, therefore, that war is not merely an act of policy but a true political instrument, a continuation of political intercourse, carried on with other means. What remains peculiar to war is simply the peculiar nature of its means. War in general, and the commander in any specific instance, is entitled to require that the trend and designs of policy shall not be inconsistent with these means. That, of course, is no small demand; but however much it may affect political aims in a given case, it will never do more than modify them. The political object is the goal, war is the means of reaching it, and means can never be considered in isolation from their purpose.

25. The Diverse Nature of War

The more powerful and inspiring the motives for war, the more they affect the belligerent nations and the fiercer the tensions that precede the outbreak, the closer will war approach its abstract concept, the more important will be the destruction of the enemy, the more closely will the military aims and the political objects of war coincide, and the more military and less political will war appear to be. On the other hand, the less intense the motives, the less will the military element's natural tendency to violence coincide with political directives. As a result, war will be driven further from its natural course, the political object will be more and more at variance with the aim of ideal war, and the conflict will seem increasingly *political* in character.

At this point, to prevent the reader from going astray, it must be observed that the phrase, the *natural tendency* of war, is used in its philosophical, strictly *logical* sense alone and does not refer to the tendencies of the forces that are actually engaged in fighting— including, for instance, the morale and emotions of the combatants. At times, it is true, these might be so aroused that the political factor would be hard put to control them. Yet such a conflict will not occur very often, for if the motivations are so powerful there must be a policy of proportionate magnitude. On the other hand, if policy is directed only toward minor objectives, the emotions of the masses will be little stirred and they will have to be stimulated rather than held back.

26. All Wars Can Be Considered Acts of Policy

It is time to return to the main theme and observe that while policy is apparently effaced in the one kind of war and yet is strongly evident in the other, both kinds are equally political. If the state is thought of as a person, and policy as the product of its brain, then among the contingencies for which the state must be prepared is a war in which every element calls for policy to be eclipsed by violence. Only if politics is regarded not as resulting from a just appreciation of affairs, but—as it conventionally is—as cautious, devious, even dishonest, shying away from force, could the second type of war appear to be more 'political' than the first.

27. The Effects of This Point of View on the Understanding of Military History and the Foundations of Theory

First, therefore, it is clear that war should never be thought of as *something autonomous* but always as an *instrument of policy*; otherwise the entire history of war would contradict us. Only this approach will enable us to penetrate the problem intelligently. *Second*, this way of looking at it will show us how wars must vary with the nature of their motives and of the situations which give rise to them.

The first, the supreme, the most far-reaching act of judgement that the statesman and commander have to make is to establish by that test the kind of war on which they are embarking; neither mistaking it for, nor trying to turn it into, something that is alien to its nature. This is the first of all strategic questions and the most comprehensive. It will be given detailed study later, in the chapter on war plans*.

It is enough, for the moment, to have reached this stage and to have established the cardinal point of view from which war and the theory of war have to be examined.

28. The Consequences for Theory

War is more than a true chameleon that slightly adapts its characteristics to the given case. As a total phenomenon its dominant tendencies always make war a paradoxical trinity—composed of primordial violence, hatred, and enmity, which are to be regarded as a blind natural force; of the play of chance and probability within which the creative spirit is free to roam; and of its element of subordination, as an instrument of policy, which makes it subject to reason alone.

The first of these three aspects mainly concerns the people; the second the commander and his army; the third the government.

The passions that are to be kindled in war must already be inherent in the people; the scope which the play of courage and talent will enjoy in the realm of probability and chance depends on the particular character of the commander and the army; but the political aims are the business of government alone.

These three tendencies are like three different codes of law, deep-rooted in their subject and yet variable in their relationship to one another. A theory that ignores any one of them or seeks to fix an arbitrary relationship between them would conflict with reality to such an extent that for this reason alone it would be totally useless.

Our task therefore is to develop a theory that maintains a balance between these three tendencies, like an object suspended between three magnets.

What lines might best be followed to achieve this difficult task will be explored in the book on the theory of war [Book Two]. At any rate, the preliminary concept of war which we have formulated casts a first ray of light on the basic structure of theory, and enables us to make an initial differentiation and identification of its major components.

CHAPTER 2

PURPOSE AND MEANS IN WAR

THE preceding chapter showed that the nature of war is complex and changeable. I now propose to inquire how its nature influences its purpose and its means.

If for a start we inquire into the objective of any particular war, which must guide military action if the political purpose is to be properly served, we find that the object of any war can vary just as much as its political purpose and its actual circumstances.

If for the moment we consider the pure concept of war, we should have to say that the political purpose of war had no connection with war itself; for if war is an act of violence meant to force the enemy to do our will its aim would have *always* and *solely* to be to overcome the enemy and disarm him. That aim is derived from the theoretical concept of war; but since many wars do actually come very close to fulfilling it, let us examine this kind of war first of all.

Later, when we are dealing with the subject of war plans, we shall investigate in greater detail what is meant by *disarming* a country. But we should at once distinguish between three things, three broad objectives, which between them cover everything: the *armed forces*, the *country*, and the *enemy's will*.

The fighting forces must be *destroyed*: that is, they must be *put in such a condition that they can no longer carry on the fight*. Whenever we use the phrase 'destruction of the enemy's forces' this alone is what we mean.

The country must be occupied; otherwise the enemy could raise fresh military forces.

Yet both these things may be done and the war, that is the animosity and the reciprocal effects of hostile elements, cannot be considered to have ended so long as the enemy's *will* has not been broken: in other words, so long as the enemy government and its allies have not been driven to ask for peace, or the population made to submit.

We may occupy a country completely, but hostilities can be renewed again in the interior, or perhaps with allied help. This of course can also happen *after* the peace treaty, but this only shows that not every war necessarily leads to a final decision and settlement. But even if hostilities should occur again, a peace treaty will always extinguish a mass of sparks that might have gone on quietly smouldering. Further, tensions are slackened, for lovers of peace (and they abound among every people under all circumstances) will then abandon any thought of further action. Be that as it may, we must always consider that with the conclusion of peace the purpose of the war has been achieved and its business is at an end.

Since of the three objectives named, it is the fighting forces that assure the safety of the country, the natural sequence would be to destroy them first, and then subdue the country. Having achieved these two goals and exploiting our own position of strength, we can bring the enemy to the peace table. As a rule, destroying the enemy's forces tends to be a gradual process, as does the ensuing subjugation of the country. Normally the one reacts on the other, in that loss of territory weakens the fighting forces; but that particular sequence of events is not essential and therefore does not always take place. Before they suffer seriously, the enemy's forces may retire to remote areas, or even withdraw to other countries. In that event, of course, most or all of the country will be occupied.

But the aim of *disarming the enemy* (the object of *war in the abstract*, the ultimate means of accomplishing the war's political purpose, which should incorporate all the rest) is in fact not always encountered in reality, and need not be fully achieved as a condition of peace. On no account should theory raise it to the level of a law. Many treaties have been concluded before one of the antagonists could be called powerless—even before the balance of power had been seriously altered. What is more, a review of actual cases shows a whole category of wars in which the very idea of *defeating the enemy* is unreal: those in which the enemy is substantially the stronger power.

The reason why the object of war that emerges in theory is sometimes inappropriate to actual conflict is that war can be of two very different kinds, a point we discussed in the first chapter. If war were what pure theory postulates, a war between states of markedly unequal strength would be absurd, and so impossible. At most, material disparity could not go beyond the amount that moral factors could replace; and social conditions being what they are in Europe today, moral forces would not go far. But wars have in fact been fought between states of *very unequal strength, for actual war is often far removed from the pure concept postulated by theory.* Inability to carry on the struggle can, in practice, be replaced by two other grounds for making peace: the first is the improbability of victory; the second is its unacceptable cost.

As we saw in the first chapter, war, if taken as a whole, is bound to move from the strict law of inherent necessity toward probabilities. The more the circumstances that gave rise to the conflict cause it to do so, the slighter will be its motives and the tensions which it occasions. And this makes it understandable how an analysis of probabilities may lead to peace itself. Not every war need be fought until one side collapses. When the motives and tensions of war are slight we can imagine that the very faintest prospect of defeat might be enough to cause one side to yield. If from the very start the other side feels that this is probable, it will obviously concentrate on bringing about *this probability* rather than take the long way round and totally defeat the enemy.

Of even greater influence on the decision to make peace is the consciousness of all the effort that has already been made and of the efforts yet to come. Since war is not an act of senseless passion but is controlled by its political object, the value of this object must determine the sacrifices to be made for it in *magnitude* and also in *duration*. Once the expenditure of effort exceeds the value of the political object, the object must be renounced and peace must follow.

We see then that if one side cannot completely disarm the other, the desire for peace on either side will rise and fall with the probability of further successes and the amount of effort these would require. If such incentives were of equal strength on both sides, the two would resolve their political disputes by meeting half way. If the incentive grows on one side, it should diminish on the other. Peace will result so long as their sum total is sufficient—though the side that feels the lesser urge for peace will naturally get the better bargain.

One point is purposely ignored for the moment—the difference that the *positive* or *negative* character of the political ends is bound to produce in practice. As we shall see, the difference is important, but at this stage we must take a broader view because the original political objects can greatly alter during the course of the war and may finally change entirely *since they are influenced by events and their probable consequences*.

The question now arises how success can be made more likely. One way, of course, is to choose objectives that will incidentally bring about the enemy's collapse—*the destruction of his armed forces and the conquest of his territory*; but neither is quite what it would be if our real object were the total defeat of the enemy. When we attack the enemy, it is one thing if we mean our first operation to be followed by others until all resistance has been broken; it is quite another if our aim is only to obtain a single victory, in order to make the enemy insecure, to impress our greater strength upon him, and to give him doubts about his future. If that is the extent of our aim, we will employ no more strength than is absolutely necessary. In the same way, conquest of territory is a different matter if the enemy's collapse is not the object. If we wish to gain total victory, then the destruction of his armed forces is the most appropriate action and the occupation of his territory only a consequence. To occupy land before his armies are defeated should be considered at best a necessary evil. If on the other hand we do not aim at destroying the opposing army, and if we are convinced that the enemy does not seek a brutal decision, but rather *fears* it, then the seizure of a lightly held or undefended province is *an advantage in itself*; and should this advantage be enough to make the enemy fear for the final outcome, it can be considered as a short cut on the road to peace.

But there is another way. It is possible to increase the likelihood of success without defeating the enemy's forces. I refer to operations that have *direct political repercussions*, that are designed in the first place to disrupt the opposing alliance, or to paralyse it, that gain us new allies, favourably affect the political scene, etc. If such operations are possible it is obvious that they can greatly improve our prospects and that they can form a much shorter route to the goal than the destruction of the opposing armies.

The second question is how to influence the enemy's expenditure of effort; in other words, how to make the war more costly to him.

The enemy's expenditure of effort consists in the *wastage of his forces*—our *destruction* of them; and in his *loss of territory*—our *conquest*.

Closer study will make it obvious that both of these factors can vary in their significance with the variation in objectives. As a rule the differences will be slight, but that should not mislead us, for in practice, when strong motives are not present, the slightest nuances often decide between the different uses of force. For the moment all that matters is to show that, given certain conditions, different ways of reaching the objective are *possible* and that they are neither *inconsistent, absurd*, nor even *mistaken*.

In addition, there are three other methods directly aimed at increasing the enemy's expenditure of effort. The first of these is *invasion*, that is *the seizure of enemy territory*; *not with the object of retaining it* but in order to exact financial contributions, or even to lay it waste. The immediate object here is neither to conquer the enemy country nor to destroy its army, but simply *to cause general damage*. The second method is to give priority to operations that will increase the enemy's suffering. It is easy to imagine two alternatives: one operation is far more advantageous if the purpose is to defeat the enemy; the other is more profitable if that cannot be done. The first tends to be described as the more military, the second the more political alternative. From the highest point of view, however, one is as military as the other, and neither is appropriate unless it suits the particular conditions. The third, and far the most important method, judging from the frequency of its use, is *to wear down* the enemy. That expression is more than a label; it describes the process precisely, and is not so metaphorical as it may seem at first. Wearing down the enemy in a conflict means using *the duration of the war to bring about a gradual exhaustion of his physical and moral resistance*.

If we intend to hold out longer than our opponent we must be content with the smallest possible objects, for obviously a major object requires more effort than a minor one. The minimum object is *pure self-defence*; in other words, fighting without a positive purpose. With such a policy our relative strength will be at its height, and thus the prospects for a favourable outcome will be greatest. But how far can this negativity be pushed? Obviously not to the point of absolute passivity, for sheer endurance would not be fighting at all. But resistance is a form of action, aimed at destroying enough of the enemy's power to force him to renounce his intentions. Every single act of our resistance is directed to that act alone, and that is what makes our policy negative.

Undoubtedly a single action, assuming it succeeds, would do less for a negative aim than it would for a positive one. But that is just the difference: the former is more likely to succeed and so to give you more security. What it lacks in immediate effectiveness it must make up for in its use of time, that is by prolonging the war. Thus the negative aim, which lies at the heart of pure resistance, is also the natural formula for outlasting the enemy, for wearing him down.

Here lies the origin of the distinction that dominates the whole of war: the difference between *attack* and *defence*. We shall not pursue the matter now, but let us just say this: that from the negative purpose derive all the advantages, all the more effective forms, of fighting, and that in it is expressed the dynamic relationship between the magnitude and the likelihood of success. All this will be gone into later.

If a negative aim—that is, the use of every means available for pure resistance—gives an advantage in war, the advantage need only be enough to *balance* any superiority the opponent may possess: in the end his political object will not seem worth the effort it costs. He must then renounce his policy. It is evident that this method, wearing down the enemy, applies to the great number of cases where the weak endeavour to resist the strong.

Frederick the Great* would never have been able to defeat Austria* in the Seven Years War:* and had he tried to fight in the manner of Charles XII* he would unfailingly have been destroyed himself. But for seven years he skilfully husbanded his strength and finally convinced the allies that far greater efforts were needed than they had foreseen. Consequently they made peace.

We can now see that in war many roads lead to success, and that they do not all involve the opponent's outright defeat. They range from *the destruction of the enemy's forces, the conquest of his territory, to a temporary occupation or invasion, to projects with an immediate political purpose, and finally to passively awaiting the enemy's attacks.* Any one of these may be used to overcome the enemy's will: the choice depends on circumstances. One further kind of action, of shortcuts to the goal, needs mention: one could call them arguments *ad hominem.** Is there a field of human affairs where personal relations do not count, where the sparks they strike do not leap across all practical considerations? The personalities of statesmen and soldiers are such important factors that in war above all it is vital not to underrate them. It is enough to mention this point: it would be pedantic to attempt a systematic classification. It can be said, however, that these questions of personality and personal relations raise the number of possible ways of achieving the goal of policy to infinity.

To think of these shortcuts as rare exceptions, or to minimize the difference they can make to the conduct of war, would be to underrate them. To avoid that error we need only bear in mind how wide a range of political interests can lead to war, or think for a moment of the gulf that separates a war of annihilation, a struggle for political existence, from a war reluctantly declared in consequence of political pressure or of an alliance that no longer seems to reflect the state's true interests. Between these two extremes lie numerous gradations. If we reject a single one of them on theoretical grounds, we may as well reject them all, and lose contact with the real world.

So much then for the ends to be pursued in war; let us now turn to the means.

There is only one: *combat.* However many forms combat takes, however far it may be removed from the brute discharge of hatred and enmity of a physical encounter, however many forces may intrude which themselves are not part of fighting, it is inherent in the very concept of war that everything that occurs *must originally derive from combat*.

It is easy to show that this is always so, however many forms reality takes. Everything that occurs in war results from the existence of armed forces; *but whenever armed forces, that is armed individuals*, are used, the idea of combat must be present.

Warfare comprises everything related to the fighting forces— everything to do with their creation, maintenance, and use.

Creation and maintenance are obviously only means; their use constitutes the end.

Combat in war is not a contest between individuals. It is a whole made up of many parts, and in that whole two elements may be distinguished, one determined by the subject, the other by the objective. The mass of combatants in an army endlessly forms fresh elements, which themselves are parts of a greater structure. The fighting activity of each of these parts constitutes a more or less clearly defined element. Moreover, combat itself is made an element of war by its very purpose, by its *objective*.

Each of these elements which become distinct in the course of fighting is named an *engagement*.

If the idea of fighting underlies every use of the fighting forces, then their employment means simply the planning and organizing of a series of engagements.

The whole of military activity must therefore relate directly or indirectly to the engagement. The end for which a soldier is recruited, clothed, armed, and trained, the whole object of his sleeping, eating, drinking, and marching *is simply that he should fight at the right place and the right time*.

If all threads of military activity lead to the engagement, then if we control the engagement, we comprehend them all. Their results are produced by our orders and by the execution of these orders, never directly by other conditions. Since in the engagement every-thing is concentrated on the destruction of the enemy, or rather of *his armed forces*, which is inherent in its very concept, it follows that the destruction of the enemy's forces is always the means by which the purpose of the engagement is achieved.

The purpose in question may be the destruction of the enemy's forces, but not necessarily so; it may be quite different. As we have shown, the destruction of the enemy is not the only means of attaining the political object, when there are other objectives for which the war is waged. It follows that those other objectives can also become the purpose of particular military operations, and thus also the purpose of engagements.

Even when subordinate engagements are directly intended to destroy the opposing forces, that destruction still need not be their first, immediate concern.

Bearing in mind the elaborate structure of an army, and the numerous factors that determine its employment, one can see that the fighting activity of such a force is also subject to complex organization, division of functions, and combinations. The separate units obviously must often be assigned tasks that are not in themselves concerned with the destruction of the enemy's forces, which may indeed increase their losses but do so only indirectly. If a battalion is ordered to drive the enemy from a hill, a bridge, etc., the true purpose is normally to occupy that point. Destruction of the enemy's force is only a means to an end, a secondary matter. If a mere demonstration is enough to cause the enemy to abandon his position, the objective has been achieved; but as a rule the hill or bridge is captured only so that even more damage can be inflicted on the enemy. If this is the case on the battlefield, it will be even more so in the theatre of operations, where it is not merely two armies that are facing each other, but two states, two peoples, two nations. The range of possible circumstances, and therefore of options, is greatly increased, as is the variety of dispositions; and the gradation of objects at various levels of command will further separate the first means from the ultimate purpose.

Thus there are many reasons why the purpose of an engagement may not be the destruction of the enemy's forces, the forces immediately confronting us. Destruction may be merely a means to some other end. In such a case, total destruction has ceased to be the point; the engagement is nothing but a *trial of strength*. In itself it is of no value; its significance lies in the outcome of the trial.

When one force is a great deal stronger than the other, an estimate may be enough. There will be no fighting: the weaker side will yield at once.

The fact that engagements do not always aim at the destruction of the opposing forces, that their objectives can often be attained without any fighting at all but merely by an evaluation of the situation, explains why entire campaigns can be conducted with great energy even though actual fighting plays an unimportant part in them.

This is demonstrated by hundreds of examples in the history of war. Here we are only concerned to show that it is *possible*; we need not ask how often it was appropriate, in other words *consistent with the overall purpose*, to avoid the test of battle, or whether all the reputations made in such campaigns would stand the test of critical examination.

There is only one means in war: combat. But the multiplicity of forms that combat assumes leads us in as many different directions as are created by the multiplicity of aims, so that our analysis does not seem to have made any progress. But that is not so: the fact that only one means exists constitutes a strand that runs through the entire web of military activity and really holds it together.

We have shown that the destruction of the enemy's forces is one of the many objects that can be pursued in war, and we have left aside the question of its importance relative to other purposes. In any given case the answer will depend on circumstances; its importance to war in general remains to be clarified. We shall now go into this question, and we shall see what value must necessarily be attributed to this object of destruction.

Combat is the only effective force in war; its aim is to destroy the enemy's forces as a means to a further end. That holds good even if no actual fighting occurs, because the outcome rests on the assumption that if it came to fighting, the enemy would be destroyed. It follows that the destruction of the enemy's force underlies all military actions; all plans are ultimately based on it, resting on it like an arch on its abutment. Consequently, all action is undertaken in the belief that if the ultimate test of arms should actually occur, the outcome would be *favourable*. The decision by arms is for all major and minor operations in war what cash payment is in commerce.* Regardless how complex the relationship between the two parties, regardless how rarely settlements actually occur, they can never be entirely absent.

If a decision by fighting is the basis of all plans and operations, it follows that the enemy *can frustrate everything through a successful battle*. This occurs not only when the encounter affects an essential factor in our plans, but when any victory that is won is of sufficient scope. For every important victory—that is, destruction of opposing forces—reacts on all other possibilities. Like liquid, they will settle at a new level.

Thus it is evident that destruction of the enemy forces is always the superior, more effective means, with which others cannot compete. But of course, we can only say destruction of the enemy is more effective if we can assume that all other conditions are equal. It would be a great mistake to deduce from this argument that a headlong rush must always triumph over skilful caution. Blind aggressiveness would destroy the attack itself, not the defence, and this is not what we are talking about. Greater effectiveness relates not to the *means* but to the *end*; we are simply comparing the effect of different outcomes.

When we speak of destroying the enemy's forces we must emphasize that nothing obliges us to limit this idea to physical forces: the moral element must also be considered. The two interact throughout: they are inseparable. We have just mentioned the effect that a great destructive act—a major victory—inevitably exerts on all other actions, and it is exactly at such times that the moral factor is, so to speak, the most fluid element of all, and therefore spreads most easily to affect everything else. The advantage that the destruction of the enemy possesses over all other means is balanced by its cost and danger; and it is only in order to avoid these risks that other policies are employed.

That the method of destruction cannot fail to be expensive is understandable; other things being equal, the more intent we are on destroying the enemy's forces, the greater our own efforts must be.

The danger of this method is that the greater the success we seek, the greater will be the damage if we fail.

Other methods, therefore, are less costly if they succeed and less damaging if they fail, though this holds true only if both sides act identically, if the enemy pursues the same course as we do. If he were to seek the decision through a major battle, *his choice would force us against our will to do likewise.* Then the outcome of the battle would be decisive; but it is clear—other things again being equal—that we would be at an overall disadvantage, since our plans and resources had been in part intended to achieve other goals, whereas the enemy's were not. Two objectives, neither of which is part of the other, are mutually exclusive: one force cannot simultaneously be used for both. If, therefore, one of the two commanders is resolved to seek a decision through major battles, he will have an excellent chance of success if he is certain that his opponent is pursuing a different policy. Conversely, the commander who wishes to adopt different means can reasonably do so only if he assumes his opponent to be equally unwilling to resort to major battles.

What has been said about plans and forces being directed to other uses refers only to the *positive* purposes, other than the destruction of enemy forces, that can be pursued in war. It pertains *in no way to pure resistance*, which seeks to wear down the opponent's strength. Pure resistance has no *positive* intention; we can use our forces only to frustrate the enemy's intentions, and not divert them to other objectives.

Here we must consider the negative side of destroying the enemy's forces—that is, the preservation of our own. These two efforts always go together; they interact. They are integral parts of a single purpose, and we only need to consider the result if one or the other dominates. The effort to destroy the enemy's forces has a positive purpose and leads to positive results, whose final aim is the enemy's collapse. Preserving our own forces has a negative purpose; it frustrates the enemy's intentions—that is, it amounts to pure resistance, whose ultimate aim can only be to prolong the war until the enemy is exhausted.

The policy with a positive purpose calls the act of destruction into being; the policy with a negative purpose waits for it.

How far such a waiting attitude may or should be maintained is a question we shall study in connection with the theory of attack and defence, whose basic element is here involved. For the moment we need only say that a policy of waiting must never become passive endurance, that any action involved in it may just as well seek the destruction of the opposing forces as any other objective. It would be a fundamental error to imagine that a negative aim implies a preference for a bloodless decision over the destruction of the enemy. A preponderantly negative effort may of course lead to such a choice, but always at the risk that it is not the appropriate course: that depends on factors that are determined not by us but by the opponent. Avoidance of bloodshed, then, should not be taken as an act of policy if our main concern is to preserve our forces. On the contrary, if such a policy did not suit the particular situation it would lead our forces to disaster. A great many generals have failed through this mistaken assumption.

The one certain effect a preponderantly negative policy will have is to retard the decision: in other words, action is transposed into waiting for the decisive moment. This usually means that *action is postponed* in time and space to the extent that space is relevant and circumstances permit. If the time arrives when further waiting would bring excessive disadvantages, then the benefit of the negative policy has been exhausted. The destruction of the enemy—an aim that has until then been postponed but not displaced by another consideration—now re-emerges.

Our discussion has shown that while in war many different roads can lead to the goal, to the attainment of the political object, fighting is the only possible means. Everything is governed by a supreme law, the *decision by force of arms*. If the opponent does seek battle, this recourse can never be denied him. A commander who prefers another strategy must first be sure that his opponent either will not appeal to that supreme tribunal—force—or that he will lose the verdict if he does. To sum up: of all the possible aims in war, the destruction of the enemy's armed forces always appears as the highest.

At a later stage and by degrees we shall see what other kinds of strategies can achieve in war. All we need to do for the moment is to admit the general *possibility of their existence*, the possibility of deviating from the basic concept of war under the pressure of special circumstances. But even at this point we must not fail to emphasize that the *violent resolution of the crisis*, the wish to annihilate the enemy's forces, is the first-born son of war. If the political aims are small, the motives slight and tensions low, a prudent general may look for any way to avoid major crises and decisive actions, exploit any weaknesses in the opponent's military and political strategy, and finally reach a peaceful settlement. If his assumptions are sound and promise success we are not entitled to criticize him. But he must never forget that he is moving on devious paths where the god of war may catch him unawares. He must always keep an eye on his opponent so that he does not, if the latter has taken up a sharp sword, approach him armed only with an ornamental rapier.

These conclusions concerning the nature of war and the function of its purposes and means; the manner in which war in practice deviates in varying degrees from its basic, rigorous concept, taking this form or that, but always remaining subject to that basic concept, as to a supreme law; all these points must be kept in mind in our subsequent analyses if we are to perceive the real connections between all aspects of war, and the true significance of each; and if we wish to avoid constantly falling into the wildest inconsistencies with reality and even with our own arguments.

CHAPTER 3

ON MILITARY GENIUS

ANY complex activity, if it is to be carried on with any degree of virtuosity, calls for appropriate gifts of intellect and temperament. If they are outstanding and reveal themselves in exceptional achievements, their possessor is called a 'genius'.

We are aware that this word is used in many senses, differing both in degree and in kind. We also know that some of these meanings make it difficult to establish the essence of genius. But since we claim no special expertise in philosophy or grammar, we may be allowed to use the word in its ordinary meaning, in which 'genius' refers to a very highly developed mental aptitude for a particular occupation.

Let us discuss this faculty, this distinction of mind for a moment, setting out its claims in greater detail, so as to gain a better understanding of the concept. But we cannot restrict our discussion to *genius* proper, as a superlative degree of talent, for this concept lacks measurable limits. What we must do is to survey all those gifts of mind and temperament that in combination bear on military activity. These, taken together, constitute *the essence of military genius*. We have said *in combination*, since it is precisely the essence of military genius that it does not consist in a single appropriate gift—courage, for example—while other qualities of mind or temperament are wanting or are not suited to war. Genius consists *in a harmonious combination of elements*, in which one or the other ability may predominate, but none may be in conflict with the rest.

If every soldier needed some degree of military genius our armies would be very weak, for the term refers to a special cast of mental or moral powers which can rarely occur in an army when a society has to employ its abilities in many different areas. The smaller the range of activities of a nation and the more the military factor dominates, the greater will be the incidence of military genius. This, however, is true only of its distribution, not of its quality. The latter depends on the *general intellectual development* of a given society. In any primitive, warlike race, the warrior spirit is far more common than among civilized peoples. It is possessed by almost every warrior: but in civilized societies only necessity will stimulate it in the people as a whole, since they lack the natural disposition for it. On the other hand, we will never find a savage who is a truly great commander, and very rarely one who would be considered a military genius, since this requires a degree of intellectual powers beyond anything that a primitive people can develop. Civilized societies, too, can obviously possess a warlike character to greater or lesser degree, and the more they develop it, the greater will be the number of men with military spirit in their armies. Possession of military genius coincides with the higher degrees of civilization: the most highly developed societies produce the most brilliant soldiers, as the Romans and the French have shown us. With them, as with every people renowned in war, the greatest names do not appear before a high level of civilization has been reached.

We can already guess how great a role intellectual powers play in the higher forms of military genius. Let us now examine the matter more closely.

War is the realm of danger; therefore *courage* is the soldier's first requirement.

Courage is of two kinds: courage in the face of personal danger, and courage to accept responsibility, either before the tribunal of some outside power or before the court of one's own conscience. Only the first kind will be discussed here.

Courage in face of personal danger is also of two kinds. It may be indifference to danger, which could be due to the individual's constitution, or to his holding life cheap, or to habit. In any case, it must be regarded as a permanent *condition*. Alternatively, courage may result from such positive motives as ambition, patriotism, or enthusiasm of any kind. In that case courage is a feeling, an emotion, not a permanent state.

These two kinds of courage act in different ways. The first is the more dependable; having become second nature, it will never fail. The other will often achieve more. There is more reliability in the first kind, more boldness in the second. The first leaves the mind calmer; the second tends to stimulate, but it can also blind. *The highest kind of courage is a compound of both.*

War is the realm of physical exertion and suffering. These will destroy us unless we can make ourselves indifferent to them, and for this birth or training must provide us with a certain strength of body and soul. If we do possess those qualities, then even if we have nothing but common sense to guide them we shall be well equipped for war: it is exactly these qualities that primitive and semicivilized peoples usually possess.

If we pursue the demands that war makes on those who practise it, we come to the region dominated by the *powers of intellect*. War is the realm of uncertainty; three quarters of the factors on which action in war is based are wrapped in a fog of greater or lesser uncertainty. A sensitive and discriminating judgement is called for; a skilled intelligence to scent out the truth.

Average intelligence may recognize the truth occasionally, and exceptional courage may now and then retrieve a blunder; but usually intellectual inadequacy will be shown up by indifferent achievement.

War is the realm of chance. No other human activity gives it greater scope: no other has such incessant and varied dealings with this intruder. Chance makes everything more uncertain and interferes with the whole course of events.

Since all information and assumptions are open to doubt, and with chance at work everywhere, the commander continually finds that things are not as he expected. This is bound to influence his plans, or at least the assumptions underlying them. If this influence is sufficiently powerful to cause a change in his plans, he must usually work out new ones; but for these the necessary information may not be immediately available. During an operation decisions have usually to be made at once: there may be no time to review the situation or even to think it through. Usually, of course, new information and re-evaluation are not enough to make us give up our intentions: they only call them in question. We now know more, but this makes us more, not less uncertain. The latest reports do not arrive all at once: they merely trickle in. They continually impinge on our decisions, and our mind must be permanently armed, so to speak, to deal with them.

If the mind is to emerge unscathed from this relentless struggle with the unforeseen, two qualities are indispensable: *first, an intellect that, even in the darkest hour, retains some glimmerings of the inner light which leads to truth; and second, the courage to follow this faint light wherever it may lead.* The first of these qualities is described by the French term, *coup d'oeil*; the second is *determination*.

The aspect of war that has always attracted the greatest attention is the engagement. Because time and space are important elements of the engagement, and were particularly significant in the days when the cavalry attack was the decisive factor, the *idea of a rapid and accurate decision* was first based on an evaluation of time and space, and consequently received a name which refers to visual estimates only. Many theorists of war have employed the term in that limited sense. But soon it was also used of any sound decision taken in the midst of action—such as recognizing the right point to attack, etc. *Coup d'oeil* therefore refers not alone to the physical but, more commonly, to the inward eye. The expression, like the quality itself, has certainly always been more applicable to tactics, but it must also have its place in strategy, since here as well quick decisions are often needed. Stripped of metaphor and of the restrictions imposed on it by the phrase, the concept merely refers to the quick recognition of a truth that the mind would ordinarily miss or would perceive only after long study and reflection.

Determination in a single instance is an expression of courage; if it becomes characteristic, a mental habit. But here we are referring not to physical courage but to the courage to accept responsibility, courage in the face of a moral danger. This has often been called *courage d'esprit*, because it is created by the intellect. That, however, does not make it an act of the intellect: it is an act of temperament. Intelligence alone is not courage; we often see that the most intelligent people are irresolute. Since in the rush of events a man is governed by feelings rather than by thought, the intellect needs to arouse the quality of courage, which then supports and sustains it in action.

Looked at in this way, the role of determination is to limit the agonies of doubt and the perils of hesitation when the motives for action are inadequate. Colloquially, to be sure, the term 'determination' also applies to a propensity for daring, pugnacity, boldness, or temerity. But when a man has adequate grounds for action—whether subjective or objective, valid or false—he cannot properly be called 'determined'. This would amount to putting oneself in his position and weighting the scale with a doubt that he never felt. In such a case it is only a question of strength or weakness. I am not such a pedant as to quarrel with common usage over a slight misuse of a word; the only purpose of these remarks is to preclude misunderstandings.

Determination, which dispels doubt, is a quality that can be aroused only by the intellect, and by a specific cast of mind at that. More is required to create determination than a mere conjunction of superior insight with the appropriate emotions. Some may bring the keenest brains to the most formidable problems, and may possess the courage to accept serious responsibilities; but when faced with a difficult situation they still find themselves unable to reach a decision. Their courage and their intellect work in separate compartments, not together; determination, therefore, does not result. It is engendered only by a *mental act*; the mind tells man that boldness is required, and thus gives direction to his will. This particular cast of mind, which employs the fear of *wavering* and *hesitating* to suppress all other fears, is the force that makes strong men determined. Men of low intelligence, therefore, cannot possess determination in the sense in which we use the word. They may act without hesitation in a crisis, but if they do, they act *without reflection*; and a man who acts without reflection cannot, of course, be torn by doubt. From time to time action of this type may even be appropriate; but, as I have said before, it is the *average result* that indicates the existence of military genius. The statement may surprise the reader who knows some determined cavalry officers who are little given to deep thought: but he must remember that we are talking about a special kind of intelligence, not about great powers of meditation.

In short, we believe that determination proceeds from a special type of mind, from a strong rather than a brilliant one. We can give further proof of this interpretation by pointing to the many examples of men who show great determination as junior officers, but lose it as they rise in rank. Conscious of the need to be decisive, they also recognize the risks entailed by a *wrong* decision; since they are unfamiliar with the problems now facing them, their mind loses its former incisiveness. The more used they had been to instant action, the more their timidity increases as they realize the dangers of the vacillation that ensnares them.

Having discussed *coup d'oeil* and determination it is natural to pass to a related subject: *presence of mind*. This must play a great role in war, the domain of the unexpected, since it is nothing but an increased capacity of dealing with the unexpected. We admire presence of mind in an apt repartee, as we admire quick thinking in the face of danger. Neither needs to be exceptional, so long as it meets the situation. A reaction following long and deep reflection may seem quite commonplace; as an immediate response, it may give keen pleasure. The expression 'presence of mind' precisely conveys the speed and immediacy of the help provided by the intellect.

Whether this splendid quality is due to a special cast of mind or to steady nerves depends on the nature of the incident, but neither can ever be entirely lacking. A quick retort shows wit; resourcefulness in sudden danger calls, above all, for steady nerve.

Four elements make up the climate of war: danger, exertion, uncertainty, and chance. If we consider them together, it becomes evident how much fortitude of mind and character are needed to make progress in these impeding elements with safety and success. According to circumstance, reporters and historians of war use such terms as *energy, firmness, staunchness, emotional balance*, and *strength of character*. These products of a heroic nature could almost be treated as one and the same force—strength of will—which adjusts itself to circumstances: but though closely linked, they are not identical. A closer study of the interplay of psychological forces at work here may be worth while.

To begin with, clear thought demands that we keep one point in mind: of the weight, the burden, the resistance—call it what you like—that challenges the psychological strength of the soldier, only a small part is the *direct result of the enemy's activity, his resistance, or his operations*. The direct and primary impact of enemy activity falls, initially, on the soldier's person without affecting him in his capacity as commander. If, for example, the enemy resists four hours instead of two, the commander is in danger twice as long; but the higher an officer's rank, the less significant this factor becomes, and to the commander-in-chief it means nothing at all.

A second way in which the enemy's resistance *directly* affects the commander is the loss that is caused by prolonged resistance and the influence this exerts on his sense of responsibility. The deep anxiety which he must experience works on his strength of will and puts it to the test. Yet we believe that this is not by any means the heaviest burden he must bear, for he is answerable to himself alone. All other effects of enemy action, however, are felt by the men under his command, and *through them react on him.*

So long as a unit fights cheerfully, with spirit and élan, great strength of will is rarely needed; but once conditions become difficult, as they must when much is at stake, things no longer run like a well-oiled machine. The machine itself begins to resist, and the commander needs tremendous will-power to overcome this resistance. The machine's *resistance* need not consist of disobedience and argument, though this occurs often enough in individual soldiers. It is the impact of the ebbing of moral and physical strength, of the heart-rending spectacle of the dead and wounded, that the commander has to withstand—first in himself, and then in all those who, directly or indirectly, have entrusted him with their thoughts and feelings, hopes and fears. As each man's strength gives out, as it no longer responds to his will, the inertia of the whole gradually comes to rest on the commander's will alone. The ardour of his spirit must rekindle the flame of purpose in all others; his inward fire must revive their hope. Only to the extent that he can do this will he retain his hold on his men and keep control. Once that hold is lost, once his own courage can no longer revive the courage of his men, the mass will drag him down to the brutish world where danger is shirked and shame is unknown. Such are the burdens in battle that the commander's courage and strength of will must overcome if he hopes to achieve outstanding success. The burdens increase with the number of men in his command, and therefore the higher his position, the greater the strength of character he needs to bear the mounting load.

Energy in action varies in proportion to the strength of its motive, whether the motive be the result of intellectual conviction or of emotion. Great strength, however, is not easily produced where there is no emotion.

Of all the passions that inspire man in battle, none, we have to admit, is so powerful and so constant as the longing for honour and renown. The German language unjustly tarnishes this by associating it with two ignoble meanings in the terms 'greed for honor' (*Ehrgeiz*) and 'hankering after glory' (*Ruhmsucht*). The abuse of these noble ambitions has certainly inflicted the most disgusting outrages on the human race; nevertheless their origins entitle them to be ranked among the most elevated in human nature. In war they act as the essential breath of life that animates the inert mass. Other emotions may be more common and more venerated—patriotism, idealism, vengeance, enthusiasm of every kind—but they are no substitute for a thirst for fame and honour. They may, indeed, rouse the mass to action and inspire it, but they cannot give the commander the ambition to strive higher than the rest, as he must if he is to distinguish himself. They cannot give him, as can ambition, a personal, almost proprietary interest in every aspect of fighting, so that he turns each opportunity to best advantage—ploughing with vigour, sowing with care, in the hope of reaping with abundance. It is primarily this spirit of endeavour on the part of commanders at all levels, this inventiveness, energy, and competitive enthusiasm, which vitalizes an army and makes it victorious. And so far as the commander-in-chief is concerned, we may well ask whether history has ever known a great general who was not ambitious; whether, indeed, such a figure is conceivable.

Staunchness indicates the will's resistance to a single blow; *endurance* refers to prolonged resistance.

Though the two terms are similar and are often used interchangeably, the difference between them is significant and unmistakable. Staunchness in face of a single blow may result from strong emotion, whereas intelligence helps sustain endurance. The longer an action lasts, the more deliberate endurance becomes, and this is one of its sources of strength.

We now turn to *strength of mind*, or of *character*, and must first ask what we mean by these terms.

Not, obviously, vehement display of feeling, or passionate temperament: that would strain the meaning of the phrase. We mean the ability to keep one's head at times of exceptional stress and violent emotion. Could strength of intellect alone account for such a faculty? We doubt it. Of course the opposite does not flow from the fact that some men of outstanding intellect do lose their self-control; it could be argued that a powerful rather than a capacious mind is what is needed. But it might be closer to the truth to assume that the faculty known as *self-control*—the gift of keeping calm even under the greatest stress—is rooted in temperament. It is itself an emotion which serves to balance the passionate feelings in strong characters without destroying them, and it is this balance alone that assures the dominance of the intellect. The counterweight we mean is simply the sense of human dignity, the noblest pride and deepest need of all: the urge *to act rationally at all times.* Therefore we would argue that a strong character is one *that will not be unbalanced by the most powerful emotions.*

If we consider how men differ in their emotional reactions, we first find a group with small capacity for being roused, usually known as 'stolid' or 'phlegmatic'.

Second, there are men who are extremely active, but whose feelings never rise above a certain level, men whom we know to be sensitive but calm.

Third, there are men whose passions are easily inflamed, in whom excitement flares up suddenly but soon burns out, like gunpowder. And finally we come to those who do not react to minor matters, who will be moved only very gradually, not suddenly, but whose emotions attain great strength and durability. These are the men whose passions are strong, deep, and concealed.

These variants are probably related to the *physical forces* operating in the human being—they are part of that dual organism we call the nervous system, one side of which is physical, the other psychological. With our slight scientific knowledge we have no business to go farther into that obscure field; it is important nonetheless to note the ways in which these various psychological combinations can affect military activity, and to find out how far one can look for great strength of character among them.

Stolid men are hard to throw off balance, but total lack of vigour cannot really be interpreted as strength of character. It cannot be denied, however, that the imperturbability of such men gives them a certain narrow usefulness in war. They are seldom strongly motivated, lack initiative and consequently are not particularly active; on the other hand they seldom make a serious mistake.

The salient point about the second group is that trifles can suddenly stir them to act, whereas great issues are likely to overwhelm them. This kind of man will gladly help an individual in need, but the misfortune of an entire people will only sadden him; they will not stimulate him to action.

In war such men show no lack of energy or balance, but they are unlikely to achieve anything significant unless they possess a *very powerful intellect* to provide the needed stimulus. But it is rare to find this type of temperament combined with a strong and independent mind.

Inflammable emotions, feelings that are easily roused, are in general of little value in practical life, and therefore of little value in war. Their impulses are strong but brief. If the energy of such men is joined to courage and ambition they will often prove most useful at a modest level of command, simply because the action controlled by junior officers is of short duration. Often a single brave decision, a burst of emotional force, will be enough. A daring assault is the work of a few minutes, while a hard-fought battle may last a day, and a campaign an entire year.

Their volatile emotions make it doubly hard for such men to preserve their balance; they often lose their heads, and nothing is worse on active service. All the same, it would be untrue to say that highly excitable minds could never be strong—that is, could never keep their balance even under the greatest strain. Why should they not have a sense of their own dignity, since as a rule they are among the finer natures? In fact, they usually have such a sense, but there is not time for it to take effect. Once the crisis is past, they tend to be ashamed of their behaviour. If training, self-awareness, and experience sooner or later teaches them how to be on guard against themselves, then in times of great excitement an internal counterweight will assert itself so that they too can draw on great strength of character.

Lastly, we come to men who are difficult to move but have strong feelings—men who are to the previous type like heat to a shower of sparks. These are the men who are best able to summon the titanic strength it takes to clear away the enormous burdens that obstruct activity in war. Their emotions move as great masses do—slowly but irresistibly.

These men are not swept away by their emotions so often as is the third group, but experience shows that they too can lose their balance and be overcome by blind passion. This can happen whenever they lack the noble pride of self-control, or whenever it is inadequate. We find this condition mostly among great men in primitive societies, where passion tends to rule for lack of intellectual discipline. Yet even among educated peoples and civilized societies men are often swept away by passion, just as in the Middle Ages poachers chained to stags were carried off into the forest.

We repeat again: strength of character does not consist solely in having powerful feelings, but in maintaining one's balance in spite of them. Even with the violence of emotion, judgement and principle must still function like a ship's compass, which records the slightest variations however rough the sea.

We say a man has strength of character, or simply has character, if he sticks to his convictions, whether these derive from his own opinions or someone else's, whether they represent principles, attitudes, sudden insights, or any other mental force. Such *firmness* cannot show itself, of course, if a man keeps changing his mind. This need not be the consequence of external influence; the cause may be the workings of his own intelligence, but this would suggest a peculiarly insecure mind. Obviously a man whose opinions are constantly changing, even though this is in response to his own reflections, would not be called a *man of character*. The term is applied only to men whose views are *stable and constant*. This may be because they are well thought-out, clear, and scarcely open to revision; or, in the case of indolent men, because such people are not in the habit of mental effort and therefore have no reason for altering their views; and finally, because a firm decision, based on fundamental principle derived from reflection, is relatively immune to changes of opinion.

With its mass of vivid impressions and the doubts which characterize all information and opinion, there is no activity like war to rob men of confidence in themselves and in others, and to divert them from their original course of action.

In the dreadful presence of suffering and danger, emotion can easily overwhelm intellectual conviction, and in this psychological fog it is so hard to form clear and complete insights that changes of view become more understandable and excusable. Action can never be based on anything firmer than instinct, a sensing of the truth. Nowhere, in consequence, are differences of opinion so acute as in war, and fresh opinions never cease to batter at one's convictions. No degree of calm can provide enough protection: new impressions are too powerful, too vivid, and always assault the emotions as well as the intellect.

Only those general principles and attitudes that result from clear and deep understanding can provide a *comprehensive* guide to action. It is to these that opinions on specific problems should be anchored. The difficulty is to hold fast to these results of contemplation in the torrent of events and new opinions. Often there is a gap between principles and actual events that cannot always be bridged by a succession of logical deductions. Then a measure of self-confidence is needed, and a degree of scepticism is also salutary. Frequently nothing short of an imperative principle will suffice, which is not part of the immediate thought-process, but dominates it: that principle is in all doubtful cases *to stick to one's first opinion and to refuse to change unless forced to do so by a clear conviction*. A strong faith in the overriding truth of tested principles is needed; the *vividness* of transient impressions must not make us forget that such truth as they contain is of a lesser stamp. By giving precedence, in case of doubt, to our earlier convictions, by holding to them stubbornly, our actions acquire that quality of steadiness and consistency which is termed strength of character.

It is evident how greatly strength of character depends on balanced temperament; most men of emotional strength and stability are therefore men of powerful character as well.

Strength of character can degenerate into *obstinacy*. The line between them is often hard to draw in a specific case; but surely it is easy to distinguish them in theory.

Obstinacy *is not an intellectual defect*; it comes from reluctance to admit that one is wrong. To impute this to the mind would be illogical, for the mind is the seat of judgement. Obstinacy *is a fault of temperament*. Stubbornness and intolerance of contradiction result from a special kind of *egotism*, which elevates above everything else *the pleasure of its autonomous intellect, to which others must bow*. It might also be called vanity, if it were not something superior: vanity is content with the appearance alone; obstinacy demands the material reality.

We would therefore argue that strength of character turns to obstinacy as soon as a man resists another point of view not from superior insight or attachment to some higher principle, but because he *objects instinctively*. Admittedly, this definition may not be of much practical use; but it will nevertheless help us avoid the interpretation that obstinacy is simply a more intense form of strong character. There is a basic difference between the two. They are closely related, but one is so far from being *a higher degree* of the other that we can even find extremely obstinate men who are too dense to have much strength of character.

So far our survey of the attributes that a great commander needs in war has been concerned with qualities in which mind and temperament work together. Now we must address ourselves to a special feature of military activity—possibly the most striking even though it is not the most important—which is not related to temperament, and involves merely the intellect. I mean the relationship between warfare and terrain.

This relationship, to begin with, is *a permanent factor*—so much so that one cannot conceive of a regular army operating except in a definite space. Second, its importance is *decisive in the highest degree*, for it affects the operations of all forces, and at times entirely alters them. Third, its influence may be felt in the *very smallest feature of the ground*, but it can also dominate *enormous areas*.

In these ways the relationship between warfare and terrain determines the peculiar character of military action. If we consider other activities connected with the soil—gardening, for example, farming, building, hydraulic engineering, mining, game-keeping, or forestry—none extends to more than a very limited area, and a working knowledge of that area is soon acquired. But a commander must submit his work to a partner, space, which he can never completely reconnoitre, and which because of the constant movement and change to which he is subject he can never really come to know. To be sure, the enemy is generally no better off; but the handicap, though shared, is still a handicap, and the man with enough talent and experience to overcome it will have a real advantage. Moreover it is only in a general sense that the difficulty is the same for both sides; in any particular case the defender usually knows the area far better than his opponent.

This problem is unique. To master it a special gift is needed, which is given the too restricted name of *a sense of locality*. It is the faculty of *quickly and accurately grasping the topography of any area* which enables a man to find his way about at any time. Obviously this is an act of the imagination. Things are perceived, of course, partly by the naked eye and partly by the mind, which fills the gaps with guesswork based on learning and experience, and thus constructs a whole out of the fragments that the eye can see; but if the whole is to be vividly present to the mind, imprinted like a picture, like a map, upon the brain, without fading or blurring in detail, *it can only be achieved by the mental gift that we call imagination*. A poet or painter may be shocked to find that his Muse dominates these activities as well: to him it might seem odd to say that a young gamekeeper needs an unusually powerful imagination in order to be competent. If so, we gladly admit that this is to apply the concept narrowly and to a modest task. But however remote the connection, his skill must still derive from this natural gift, for if imagination is entirely lacking it would be difficult to combine details into a clear, coherent image. We also admit that a good memory can be a great help; but are we then to think of memory as a separate gift of the mind, or does imagination, after all, imprint those pictures in the memory more clearly? The question must be left unanswered, especially since it seems difficult even to conceive of these two forces as operating separately.

That practice and a trained mind have much to do with it is undeniable. Puységur,* the celebrated quarter-master-general of Marshal Luxembourg,* writes that at the beginning of his career he had little faith in his sense of locality; when he had to ride any distance at all to get the password, he invariably lost his way.

Scope for this talent naturally grows with increased authority. A hussar or scout leading a patrol must find his way easily among the roads and tracks. All he needs are a few landmarks and some modest powers of observation and imagination. A commander-in-chief, on the other hand, must aim at acquiring an overall knowledge of the configuration of a province, of an entire country. His mind must hold a vivid picture of the road-network, the river-lines and the mountain ranges, without ever losing a sense of his immediate surroundings. Of course he can draw general information from reports of all kinds, from maps, books, and memoirs. Details will be furnished by his staff. Nevertheless it is true that with a quick, unerring sense of locality his dispositions will be more rapid and assured; he will run less risk of a certain awkwardness in his concepts, and be less dependent on others.

We attribute this ability to the imagination; but that is about the only service that war can demand from this frivolous goddess, who in most military affairs is liable to do more harm than good.

With this, we believe, we have reached the end of our review of the intellectual and moral powers that human nature needs to draw upon in war. The vital contribution of intelligence is clear throughout. No wonder then, that war, though it may appear to be uncomplicated, cannot be waged with distinction except by men of outstanding intellect.

Once this view is adopted, there is no longer any need to think that it takes a great intellectual effort to outflank an enemy position (an obvious move, performed innumerable times) or to carry out a multitude of similar operations.

It is true that we normally regard the plain, efficient soldier as the very opposite of the contemplative scholar, or of the inventive intellectual with his dazzling range of knowledge. This antithesis is not entirely unrealistic; but it does not prove that courage alone will make an efficient soldier, or that having brains and using them is not a necessary part of being a good fighting man. Once again we must insist: no case is more common than that of the officer whose energy declines as he rises in rank and fills positions that are beyond his abilities. But we must also remind the reader that outstanding effort, the kind that gives men a distinguished name, is what we have in mind. Every level of command has its own intellectual standards, its own prerequisites for fame and honour.

A major gulf exists between a commander-in-chief—a general who leads the army as a whole or commands in a theatre of operations—and the senior generals immediately subordinate to him. The reason is simple: the second level is subjected to much closer control and supervision, and thus gives far less scope for independent thought. People therefore often think outstanding intellectual ability is called for only at the top, and that for all other duties common intelligence will suffice. A general of lesser responsibility, an officer grown grey in the service, his mind well-blinkered by long years of routine, may often be considered to have developed a certain stodginess; his gallantry is respected, but his simplemindedness makes us smile. We do not intend to champion and promote these good men; it would contribute nothing to their efficiency, and little to their happiness. We only wish to show things as they are, so that the reader should not think that a brave but brainless fighter can do anything of outstanding significance in war.

Since in our view even junior positions of command require outstanding intellectual qualities for outstanding achievement, and since the standard rises with every step, it follows that we recognize the abilities that are needed if the second positions in an army are to be filled with distinction. Such officers may appear to be rather simple compared to the polymath scholar, the far-ranging business executive, the statesman; but we should not dismiss the value of their practical intelligence. It sometimes happens of course that someone who made his reputation in one rank carries it with him when he is promoted, without really deserving to. If not much is demanded of him, and he can avoid exposing his incompetence, it is difficult to decide what reputation he really deserves. Such cases often cause one to hold in low estimate soldiers who in less responsible positions might do excellent work.

Appropriate talent is needed at all levels if distinguished service is to be performed. But history and posterity reserve the name of 'genius' for those who have excelled in the highest positions—as commanders-in-chief—since here the demands for intellectual and moral powers are vastly greater.

To bring a war, or one of its campaigns, to a successful close requires a thorough grasp of national policy. On that level strategy and policy coalesce: the commander-in-chief is simultaneously a statesman.

Charles XII of Sweden is not thought of as a great genius, for he could never subordinate his military gifts to superior insights and wisdom, and could never achieve a great goal with them. Nor do we think of Henry IV* of France in this manner: he was killed before his skill in war could affect the relations between states. Death denied him the chance to prove his talents in this higher sphere, where noble feelings and a generous disposition, which effectively appeased internal dissension, would have had to face a more intractable opponent.

The great range of business that a supreme commander must swiftly absorb and accurately evaluate has been indicated in the first chapter. We argue that a commander-in-chief must also be a statesman, but he must not cease to be a general. On the one hand, he is aware of the entire political situation; on the other, he knows exactly how much he can achieve with the means at his disposal.

Circumstances vary so enormously in war, and are so indefinable, that a vast array of factors has to be appreciated—mostly in the light of probabilities alone. The man responsible for evaluating the whole must bring to his task the quality of intuition that perceives the truth at every point. Otherwise a chaos of opinions and considerations would arise, and fatally entangle judgement. Bonaparte rightly said in this connection that many of the decisions faced by the commander-in-chief resemble mathematical problems worthy of the gifts of a *Newton** or an *Euler*.*

What this task requires in the way of higher intellectual gifts is a

sense of unity and a power of judgement raised to a marvellous pitch
of vision, which easily grasps and dismisses a thousand remote possibilities which an ordinary mind would labour to identify and wear
itself out in so doing. Yet even that superb display of divination, the
sovereign eye of genius itself, would still fall short of historical significance without the qualities of character and temperament we
have described.

Truth in itself is rarely sufficient to make men act. Hence the step
is always long from cognition to volition, from knowledge to ability.
The most powerful springs of action in men lie in his emotions. He
derives his most vigorous support, if we may use the term, from that
blend of brains and temperament which we have learned to recognize in the qualities of determination, firmness, staunchness, and
strength of character.

Naturally enough, if the commander's superior intellect and
strength of character did not express themselves in the final success
of his work, and were only taken on trust, they would rarely achieve
historical importance.

What the layman gets to know of the course of military events is
usually nondescript. One action resembles another, and from a mere
recital of events it would be impossible to guess what obstacles were
faced and overcome. Only now and then, in the memoirs of generals
or of their confidants, or as the result of close historical study, are
some of the countless threads of the tapestry revealed. Most of the
arguments and clashes of opinion that precede a major operation are
deliberately concealed because they touch political interests, or
they are simply forgotten, being considered as scaffolding to be
demolished when the building is complete.

Finally, and without wishing to risk a closer definition of the
higher reaches of the spirit, let us assert that the human mind (in the
normal meaning of the term) is far from uniform. If we then ask
what sort of mind is likeliest to display the qualities of military
genius, experience and observation will both tell us that it is the
inquiring rather than the creative mind, the comprehensive rather
than the specialized approach, the calm rather than the excitable
head to which in war we would choose to entrust the fate of our
brothers and children, and the safety and honour of our country.

CHAPTER 4

ON DANGER IN WAR

To someone who has never experienced danger, the idea is attractive rather than alarming. You charge the enemy, ignoring bullets and casualties, in a surge of excitement. Blindly you hurl yourself toward icy death, not knowing whether you or anyone else will escape him. Before you lies that golden prize, victory, the fruit that quenches the thirst of ambition. Can that be so difficult? No, and it will seem even less difficult than it is. But such moments are rare; and even they are not, as is commonly thought, brief like a heartbeat, but come rather like a medicine, in recurring doses, the taste diluted by time.

Let us accompany a novice to the battlefield. As we approach the rumble of guns grows louder and alternates with the whir of cannonballs, which begin to attract his attention. Shots begin to strike close around us. We hurry up the slope where the commanding general is stationed with his large staff. Here cannonballs and bursting shells are frequent, and life begins to seem more serious than the young man had imagined. Suddenly someone you know is wounded; then a shell falls among the staff. You notice that some of the officers act a little oddly; you yourself are not as steady and collected as you were: even the bravest can become slightly distracted. Now we enter the battle raging before us, still almost like a spectacle, and join the nearest divisional commander. Shot is falling like hail, and the thunder of our own guns adds to the din. Forward to the brigadier, a soldier of acknowledged bravery, but he is careful to take cover behind a rise, a house or a clump of trees. A noise is heard that is a certain indication of increasing danger—the rattling of grapeshot on roofs and on the ground. Cannonballs tear past, whizzing in all directions, and musketballs begin to whistle around us. A little further we reach the firing line, where the infantry endures the hammering for hours with incredible steadfastness. The air is filled with hissing bullets that sound like a sharp crack if they pass close to one's head. For a final shock, the sight of men being killed and mutilated moves our pounding hearts to awe and pity.

The novice cannot pass through these layers of increasing intensity of danger without sensing that here ideas are governed by other factors, that the light of reason is refracted in a manner quite different from that which is normal in academic speculation. It is an exceptional man who keeps his powers of quick decision intact if he has never been through this experience before. It is true that (with habit) as we become accustomed to it the impression soon wears off, and in half-an-hour we hardly notice our surroundings any more; yet the ordinary man can never achieve a state of perfect unconcern in which his mind can work with normal flexibility. Here again we recognize that ordinary qualities are not enough; and the greater the area of responsibility, the truer this assertion becomes. Headlong, dogged, or innate courage, overmastering ambition, or long familiarity with danger—all must be present to a considerable degree if action in this debilitating element is not to fall short of achievements that in the study would appear as nothing out of the ordinary.

Danger is part of the friction of war. Without an accurate conception of danger we cannot understand war. That is why I have dealt with it here.

CHAPTER 5

ON PHYSICAL EFFORT IN WAR

IF no one had the right to give his views on military operations except when he is frozen, or faint from heat and thirst, or depressed from privation and fatigue, objective and accurate views would be even rarer than they are. But they would at least be subjectively valid, for the speaker's experience would precisely determine his judgement. This is clear enough when we observe in what a deprecatory, even mean and petty way men talk about the failure of some operation that they have witnessed, and even more if they actually took part. We consider that this indicates how much influence physical effort exerts, and shows how much allowance has to be made for it in all our assessments.

Among the many factors in war that cannot be measured, physical effort is the most important. Unless it is wasted, physical effort is a coefficient of all forces, and its exact limit cannot be determined. But it is significant that, just as it takes a powerful archer to bend the bow beyond the average, so it takes a powerful mind to drive his army to the limit. It is one thing for an army that has been badly defeated, is beset by danger on all sides, and is disintegrating like crumbling masonry, to seek its safety in utmost physical effort. It is altogether different when a victorious army, buoyed up by its own exhilaration, remains a willing instrument in the hands of its commander. The same effort, which in the former case can at most arouse sympathy, must be admired in the other, where it is much harder to maintain.

The inexperienced observer now comes to recognize one of the elements that seem to chain the spirit and secretly wear away men's energies.

Although we are dealing only with the efforts that a general can demand of his troops, a commander of his subordinates, in other words although we are concerned with the courage it takes to make the demand and the skill to keep up the response, we must not forget the physical exertion required of the commander himself. Since we have pursued our analysis of war conscientiously to this point, we must deal with this residue as well.

Our reason for dealing with physical effort here is that like danger it is one of the great sources of friction in war. Because its limits are uncertain, it resembles one of those substances whose elasticity makes the degree of its friction exceedingly hard to gauge.

To prevent these reflections, this assessment of the impeding conditions of war, from being misused, we have a natural guide in our sensibilities. No one can count on sympathy if he accepts an insult or mistreatment because he claims to be physically handicapped. But if he manages to defend or revenge himself, a reference to his handicap will be to his advantage. In the same way, a general and an army cannot remove the stain of defeat by explaining the dangers, hardships, and exertions that were endured; but to depict them adds immensely to the credit of a victory. We are prevented from making an apparently justified statement by *our feelings*, which themselves act as a higher judgement.

CHAPTER 6

INTELLIGENCE IN WAR

BY 'intelligence' we mean every sort of information about the enemy and his country—the basis, in short, of our own plans and operations. If we consider the actual basis of this information, how unreliable and transient it is, we soon realize that war is a flimsy structure that can easily collapse and bury us in its ruins. The textbooks agree, of course, that we should only believe reliable intelligence, and should never cease to be suspicious, but what is the use of such feeble maxims? They belong to that wisdom which for want of anything better scribblers of systems and compendia resort to when they run out of ideas.

Many intelligence reports in war are contradictory; even more are false, and most are uncertain. What one can reasonably ask of an officer is that he should possess a standard of judgement, which he can gain only from knowledge of men and affairs and from common sense. He should be guided by the laws of probability. These are difficult enough to apply when plans are drafted in an office, far from the sphere of action; the task becomes infinitely harder in the thick of fighting itself, with reports streaming in. At such times one is lucky if their contradictions cancel each other out, and leave a kind of balance to be critically assessed. It is much worse for the novice if chance does not help him in that way, and on the contrary one report tallies with another, confirms it, magnifies it, lends it colour, till he has to make a quick decision—which is soon recognized to be mistaken, just as the reports turn out to be lies, exaggerations, errors, and so on. In short, most intelligence is false, and the effect of fear is to multiply lies and inaccuracies. As a rule most men would rather believe bad news than good, and rather tend to exaggerate the bad news. The dangers that are reported may soon, like waves, subside; but like waves they keep recurring, without apparent reason. The commander must trust his judgement and stand like a rock on which the waves break in vain. It is not an easy thing to do. If he does not have a buoyant disposition, if experience of war has not trained him and matured his judgement, he had better make it a rule to suppress his personal convictions, and give his hopes and not his fears the benefit of the doubt. Only thus can he preserve a proper balance.

This difficulty of *accurate recognition* constitutes one of the most serious sources of friction in war, by making things appear entirely different from what one had expected. The senses make a more vivid impression on the mind than systematic thought—so much so that I doubt if a commander ever launched an operation of any magnitude without being forced to repress new misgivings from the start. Ordinary men, who normally follow the initiative of others, tend to lose self-confidence when they reach the scene of action: things are not what they expected, the more so as they still let others influence them. But even the man who planned the operation and now sees it being carried out may well lose confidence in his earlier judgement; whereas self-reliance is his best defence against the pressures of the moment. War has a way of masking the stage with scenery crudely daubed with fearsome apparitions. Once this is cleared away, and the horizon becomes unobstructed, developments will confirm his earlier convictions—this is one of the great chasms between *planning and execution.*

CHAPTER 7

FRICTION IN WAR

IF one has never personally experienced war, one cannot understand in what the difficulties constantly mentioned really consist, nor why a commander should need any brilliance and exceptional ability. Everything looks simple; the knowledge required does not look remarkable, the strategic options are so obvious that by comparison the simplest problem of higher mathematics has an impressive scientific dignity. Once war has actually been seen the difficulties become clear; but it is still extremely hard to describe the unseen, all-pervading element that brings about this change of perspective.

Everything in war is very simple, but the simplest thing is difficult. The difficulties accumulate and end by producing a kind of friction that is inconceivable unless one has experienced war. Imagine a traveller who late in the day decides to cover two more stages before nightfall. Only four or five hours more, on a paved highway with relays of horses: it should be an easy trip. But at the next station he finds no fresh horses, or only poor ones; the country grows hilly, the road bad, night falls, and finally after many difficulties he is only too glad to reach a resting place with any kind of primitive accommodation. It is much the same in war. Countless minor incidents—the kind you can never really foresee—combine to lower the general level of performance, so that one always falls far short of the intended goal. Iron will-power can overcome this friction; it pulverizes every obstacle, but of course it wears down the machine as well. We shall often return to this point. The proud spirit's firm will dominates the art of war as an obelisk dominates the town square on which all roads converge.

Friction is the only concept that more or less corresponds to the factors that distinguish real war from war on paper. The military machine—the army and everything related to it—is basically very simple and therefore seems easy to manage. But we should bear in mind that none of its components is of one piece: each part is composed of individuals, every one of whom retains his potential of friction. In theory it sounds reasonable enough: a battalion commander's duty is to carry out his orders; discipline welds the battalion together, its commander must be a man of tested capacity, and so the great beam turns on its iron pivot with a minimum of friction. In fact, it is different, and every fault and exaggeration of the theory is instantly exposed in war. A battalion is made up of individuals, the least important of whom may chance to delay things or somehow make them go wrong. The dangers inseparable from war and the physical exertions war demands can aggravate the problem to such an extent that they must be ranked among its principal causes.

This tremendous friction, which cannot, as in mechanics, be reduced to a few points, is everywhere in contact with chance, and brings about effects that cannot be measured, just because they are largely due to chance. One, for example, is the weather. Fog can prevent the enemy from being seen in time, a gun from firing when it should, a report from reaching the commanding officer. Rain can prevent a battalion from arriving, make another late by keeping it not three but eight hours on the march, ruin a cavalry charge by bogging the horses down in mud, etc.

We give these examples simply for illustration, to help the reader follow the argument. It would take volumes to cover all difficulties. We could exhaust the reader with illustrations alone if we really tried to deal with the whole range of minor troubles that must be faced in war. The few we have given will be excused by those readers who have long since understood what we are after.

Action in war is like movement in a resistant element. Just as the simplest and most natural of movements, walking, cannot easily be performed in water, so in war it is difficult for normal efforts to achieve even moderate results. A genuine theorist is like a swimming teacher, who makes his pupils practise motions on land that are meant to be performed in water. To those who are not thinking of swimming the motions will appear grotesque and exaggerated. By the same token, theorists who have never swum, or who have not learned to generalize from experience, are impractical and even ridiculous: they teach only what is already common knowledge: how to walk.

Moreover, every war is rich in unique episodes. Each is an uncharted sea, full of reefs. The commander may suspect the reefs' existence without ever having seen them; now he has to steer past them in the dark. If a contrary wind springs up, if some major mischance appears, he will need the greatest skill and personal exertion, and the utmost presence of mind, though from a distance everything may seem to be proceeding automatically. An understanding of friction is a large part of that much-admired sense of warfare which a good general is supposed to possess. To be sure, the best general is not the one who is most familiar with the idea of friction, and who takes it most to heart (he belongs to the anxious type so common among experienced commanders). The good general must know friction in order to overcome it whenever possible, and in order not to expect a standard of achievement in his operations which this very friction makes impossible. Incidentally, it is a force that theory can never quite define. Even if it could, the development of instinct and tact would still be needed, a form of judgement much more necessary in an area littered by endless minor obstacles than in great, momentous questions, which are settled in solitary deliberation or in discussion with others. As with a man of the world instinct becomes almost habit so that he always acts, speaks, and moves appropriately, so only the experienced officer will make the right decision in major and minor matters—at every pulse-beat of war. Practice and experience dictate the answer: 'this is possible, that is not.' So he rarely makes a serious mistake, such as can, in war, shatter confidence and become extremely dangerous if it occurs often.

Friction, as we choose to call it, is the force that makes the apparently easy so difficult. We shall frequently revert to this subject, and it will become evident that an eminent commander needs more than experience and a strong will. He must have other exceptional abilities as well.

CHAPTER 8

CONCLUDING OBSERVATIONS ON BOOK ONE

WE have identified danger, physical exertion, intelligence, and friction as the elements that coalesce to form the atmosphere of war, and turn it into a medium that impedes activity. In their restrictive effects they can be grouped into a single concept of general friction. Is there any lubricant that will reduce this abrasion? Only one, and a commander and his army will not always have it readily available: combat experience.

Habit hardens the body for great exertions, strengthens the heart in great peril, and fortifies judgement against first impressions. Habit breeds that priceless quality, calm, which, passing from hussar and rifleman up to the general himself, will lighten the commander's task.

In war the experienced soldier reacts rather in the same way as the human eye does in the dark: the pupil expands to admit what little light there is, discerning objects by degrees, and finally seeing them distinctly. By contrast, the novice is plunged into the deepest night.

No general can accustom an army to war. Peacetime manoeuvres are a feeble substitute for the real thing; but even they can give an army an advantage over others whose training is confined to routine, mechanical drill. To plan manoeuvres so that some of the elements of friction are involved, which will train officers' judgement, common sense, and resolution is far more worthwhile than inexperienced people might think. It is immensely important that no soldier, whatever his rank, should wait for war to expose him to those aspects of active service that amaze and confuse him when he first comes across them. If he has met them even once before, they will begin to be familiar to him. This is true even of physical effort. Exertions must be practised, and the mind must be made even more familiar with them than the body. When exceptional efforts are required of him in war, the recruit is apt to think that they result from mistakes, miscalculations, and confusion at the top. In consequence, his morale is doubly depressed. If manoeuvres prepare him for exertions, this will not occur.

Another very useful, though more limited, way of gaining familiarity with war in peacetime is to attract foreign officers who have seen active service. Peace does not often reign everywhere in Europe, and never throughout the whole world. A state that has been at peace for many years should try to attract some experienced officers—only those, of course, who have distinguished themselves. Alternatively, some of its own officers should be sent to observe operations, and learn what war is like.

However few such officers may be in proportion to an army, their influence can be very real. Their experience, their insights, and the maturity of their character will affect their subordinates and brother officers. Even when they cannot be given high command they should be considered as guides who know the country and can be consulted in specific eventualities.

BOOK TWO
ON THE THEORY OF WAR

CLASSIFICATIONS OF THE ART OF WAR

ESSENTIALLY war is fighting, for fighting is the only effective principle in the manifold activities generally designated as war. Fighting, in turn, is a trial of moral and physical forces through the medium of the latter. Naturally moral strength must not be excluded, for psychological forces exert a decisive influence on the elements involved in war.

The need to fight quickly led man to invent appropriate devices to gain advantages in combat, and these brought about great changes in the forms of fighting. Still, no matter how it is constituted, the concept of fighting remains unchanged. That is what we mean by war.

The first inventions consisted of weapons and equipment for the individual warrior. They have to be produced and tested before war begins; they suit the nature of the fighting, which in turn determines their design. Obviously, however, this activity must be distinguished from fighting proper; it is only the preparation for it, not its conduct. It is clear that weapons and equipment are not essential to the concept of fighting, since even wrestling is fighting of a kind.

Fighting has determined the nature of the weapons employed. These in turn influence the combat; thus an interaction exists between the two.

But fighting itself still remains a distinct activity; the more so as it operates in a peculiar element—that of danger.

Thus, if there was ever a need to distinguish between two activities, we find it here. In order to indicate the practical importance of this idea, we would suggest how often it is that the ablest man in one area is shown up as the most useless pedant in another.

In fact, it is not at all difficult to consider these two activities separately if one accepts the idea of an armed and equipped fighting force as *given*: a means about which one does not need to know anything except its chief effects in order to use it properly.

Essentially, then, the art of war is the art of using the given means in combat; there is no better term for it than the *conduct of war*. To be sure in its wider sense the art of war includes all activities that exist for the sake of war, such as the creation of the fighting forces, their raising, armament, equipment, and training.

It is essential to the validity of a theory to distinguish between these two activities. It is easy to see that if the art of war were always to start with raising armed forces and adapting them to the requirements of the particular case, it would be applicable only to those few instances where the forces available exactly matched the need. If, on the other hand, one wants a theory that is valid for the great majority of cases and not completely unsuitable for any, it must be based on the most prevalent means and their most significant effects.

The conduct of war, then, consists in the planning and conduct of fighting. If fighting consisted of a single act, no further subdivision would be needed. However, it consists of a greater or lesser number of single *acts, each complete in itself*, which, as we pointed out in Chapter 1 of Book I,* are called 'engagements' and which form new entities. This gives rise to the completely different activity of *planning and executing these engagements themselves*, and of *coordinating* each of them with the others in order to further the object of the war. One has been called *tactics*, and the other *strategy*.

The distinction between tactics and strategy is now almost universal, and everyone knows fairly well where each particular factor belongs without clearly understanding why. Whenever such categories are blindly used, there must be a deep-seated reason for it. We have tried to discover the distinction, and have to say that it was just this common usage that led to it. We reject, on the other hand, the artificial definitions of certain writers, since they find no reflection in general usage.

According to our classification, then, tactics teaches *the use of armed forces in the engagement*; strategy, *the use of engagements for the object of the war*.

The concept of a single or a self-contained engagement and the conditions on which its unity depends can be more accurately defined only when we examine it more closely. For the moment, it is enough to say that in terms of space (that is, of simultaneous engagements) its unity is bounded by the range of *personal command*. In terms of time, however (that is, of a close succession of engagements) it lasts until the turning point, which is characteristic of all engagements, has been passed.

There may be doubtful cases—those, for instance, in which a

number of engagements could perhaps also be regarded as a single one. But that will not spoil our basis for classification, since the point is common to all practical systems of classification where distinctions gradually merge on a descending scale. Thus there may be individual acts which, without a shift in point of view, may belong either to strategy or to tactics; for instance, very extended positions that are little more than a chain of posts, or arrangements for certain river-crossings.

Our classification applies to and exhausts *only the utilization of the fighting forces*. But war is served by many activities that are quite different from it; some closely related, others far removed. All these activities concern the *maintenance of the fighting forces*. While their creation and training precedes their use, maintenance is concurrent with and a necessary condition for it. Strictly speaking, however, all these should be considered as activities preparatory to battle, of the type that are so closely related to the action that they are part of military operations and alternate with actual *utilization*. So one is justified in excluding these as well as all other preparatory activities from the narrower meaning of the art of war—the actual conduct of war. Indeed, it is necessary to do this if theory is to serve its principal purpose of *discriminating between dissimilar elements*. One would not want to consider the whole business of maintenance and administration as part of the *actual conduct of war*. While it may be in constant interaction with the utilization of the troops, the two are essentially very different.

In the third chapter of Book I we pointed out that, if combat or the engagement *is defined* as the only directly effective activity, the threads of all other activities will be included because they all lead to combat. The statement meant that all these activities are thus provided with a purpose, which they will have to pursue in accordance with their individual laws. Let us elaborate further on this subject.

Activities that exist in addition to the engagement differ widely.

Some of these are in one respect part of combat proper and identical with it, while in another respect they serve to maintain the fighting forces. Others are related to maintenance alone; which has an effect on combat only because of its interaction with the outcome of the fighting.

The matters that in one respect are still part of the combat are *marches, camps, and billets*: each concerns a separate phase of existence of the troops, and when one thinks of troops, the idea of the engagement must always be present.

The rest, concerned with maintenance alone, consists of *supply, medical services, and maintenance of arms and equipment.*

Marches are completely identical with the utilization of troops. *Marching in the course of an engagement* (usually known as 'deployment')* while not entailing the actual use of weapons, is so closely and inescapably linked with it as to be an integral part of what is considered an engagement. A march that is not undertaken in the course of an engagement is simply the execution of a strategic plan. The latter determines *when, where* and *with what forces* an engagement is to be fought. The march is only the means of carrying out this plan.

A march that is not part of an engagement is thus a tool of strategy, but it is not a matter of strategy exclusively. Since the forces undertaking it may at any time become involved in an engagement, the execution of the march is subject to the laws of both tactics and strategy. If a column is ordered to take a route on the near side of a river or a range of hills, that is a strategic measure: it implies that if an engagement has to be fought in the course of the march, one prefers to offer it on the near rather than the far side.

If on the other hand a column takes a route along a ridge instead of following the road through a valley, or breaks up into several smaller columns for the sake of convenience, these are tactical measures: they concern the *manner* in which the forces are to be used in the event of an engagement.

The internal order of march bears a constant relationship to readiness for combat and is therefore of a tactical nature: it is nothing more than the first preliminary disposition for a possible engagement.

The march is the tool by which strategy deploys its effective elements, the engagements. But these often become apparent only in their effect, and not in their actual course. Inevitably, therefore, in discussion the tool has often been confused with the effective element. One speaks of decisive skilful marches, and really means the combinations of engagements to which they lead. This substitution of concept is too natural, and the brevity of expression too desirable, to call for change. But it is only a telescoped chain of ideas, and one must keep the proper meaning in mind to avoid errors.

One such error occurs when strategic combinations are believed to have a value irrespective of their tactical results. One works out marches and manoeuvres, achieves one's objective without fighting an engagement, and then deduces that it is possible to defeat the enemy without fighting. Only at a later stage shall we be able to show the immense implications of this mistake.

Although marching can be seen as an integral part of combat, it has certain aspects that do not belong here, and that therefore are neither tactical nor strategic. These include all measures taken solely for the convenience of the troops, such as building roads and bridges, and so forth. These are merely preconditions; under certain circumstances they may be closely linked with the use of troops and be virtually identical with them—for instance, when a bridge is built in full view of the enemy. But essentially these activities are alien to the conduct of war, and the theory of the latter does not cover them.

The term 'camp' is a term for any concentration of troops in readiness for action, as distinct from 'billets'. Camps are places for rest and recuperation, but they also imply strategic willingness to fight wherever they may be. But their siting does determine the engagement's basic lines—a precondition of all defensive engagements. So they are essential parts both of strategy and of tactics.

Camps are replaced by billets whenever troops are thought to need more extensive recuperation. Like camps, they are therefore strategic in location and extent, and tactical in their internal organization which is geared to readiness for action.

As a rule, of course, camps and billets serve a purpose besides that of resting the troops; they may, for instance, serve to protect a certain area or maintain a position. But their purpose may simply be to rest the troops. We have to remember that strategy may pursue a wide variety of objectives: anything that seems to offer an advantage can be the purpose of an engagement, and the maintenance of the instrument of war will often itself become the object of a particular strategic combination.

So in a case where strategy merely aims at preserving the troops, we need not have strayed far afield: the use of troops is still the main concern, since that is the point of their disposition anywhere in the theatre of war.

On the other hand, the maintenance of troops in camps or billets may call for activities that do not constitute a use of the fighting forces, such as the building of shelters, the pitching of tents, and supply and sanitary services. These are neither tactical nor strategic in nature.

Even entrenchments, where site and preparation are obviously part of the order of battle and therefore tactical, are not part of the conduct of war so far as *their actual construction* is concerned. On the contrary, troops must be taught the necessary skills and knowledge as part of their training, and the theory of combat takes all that for granted.

Of the items wholly unconnected with engagements, serving only to maintain the forces, supply is the one which most directly affects the fighting. It takes place almost every day and affects every individual. Thus it thoroughly permeates the strategic aspects of all military action. The reason why we mention the strategic aspect is that in the course of a given engagement supply will rarely tend to cause an alteration of plans—though such a change remains perfectly possible. Interaction therefore will be most frequent between strategy and matters of supply, and nothing is more common than to find considerations of supply affecting the strategic lines of a campaign and a war. Still, no matter how frequent and decisive these considerations may be, the business of supplying the troops remains an activity essentially separate from their use; its influence shows in its results alone.

The other administrative functions we have mentioned are even further removed from the use of troops. Medical services, though they are vital to an army's welfare, affect it only through a small portion of its men, and therefore exert only a weak and indirect influence on the utilization of the rest. Maintenance of equipment, other than as a constant function of the fighting forces, takes place only periodically, and will therefore rarely be taken into account in strategic calculations.

At this point we must guard against a misunderstanding. In any individual case these things may indeed be of decisive importance. The distance of hospitals and supply depots may easily figure as the sole reason for very important strategic decisions—a fact we do not want to deny or minimize. However, we are not concerned with the actual circumstances of any individual case, but with pure theory. Our contention therefore is that this type of influence occurs so rarely that we should not give the theory of medical services and replacement of munitions any serious weight in the theory of the conduct of war. Unlike the supplying of the troops, therefore, it would not seem worth while to incorporate the various ways and systems those theories might suggest, and their results, into the theory of the conduct of war.

To sum up: we clearly see that the activities characteristic of war may be split into two main categories: those *that are merely preparations for war*, and *war proper*. The same distinction must be made in theory as well.

The knowledge and skills involved in the preparations will be concerned with the creation, training and maintenance of the fighting forces. It is immaterial what label we give them, but they obviously must include such matters as artillery, fortification, so-called elementary tactics, as well as all the organization and administration of the fighting forces and the like. The theory of war proper, on the other hand, is concerned with the use of these means, once they have been developed, for the purposes of the war. All that it requires from the first group is the end product, an understanding of their main characteristics. That is what we call 'the art of war' in a narrower sense, or 'the theory of the conduct of war', or 'the theory of the use of the fighting forces'. For our purposes, they all mean the same thing.

That narrower theory, then, deals with the engagement, with fighting itself, and treats such matters as marches, camps, and billets as conditions that may be more or less identical with it. It does not comprise questions of supply, but will take these into account on the same basis *as other given factors*.

The art of war in the narrower sense must now in its turn be broken down into tactics and strategy. The first is concerned with the form of the individual engagement, the second with its use. Both affect the conduct of marches, camps, and billets only through the engagement; they become tactical or strategic questions insofar as they concern either the engagement's form or its significance.

Many readers no doubt will consider it superfluous to make such a careful distinction between two things so closely related as tactics and strategy, because they do not directly affect the conduct of operations. Admittedly only the rankest pedant would expect theoretical distinctions to show direct results on the battlefield.

The primary purpose of any theory is to clarify concepts and ideas that have become, as it were, confused and entangled. Not until terms and concepts have been defined can one hope to make any progress in examining the question clearly and simply and expect the reader to share one's views. Tactics and strategy are two activities that permeate one another in time and space but are nevertheless essentially different. Their inherent laws and mutual relationship cannot be understood without a total comprehension of both.

Anyone for whom all this is meaningless either will admit no theoretical analysis at all, or his intelligence has never been insulted by the confused and confusing welter of ideas that one so often hears and reads on the subject of the conduct of war. These have no fixed point of view; they lead to no satisfactory conclusion; they appear sometimes banal, sometimes absurd, sometimes simply adrift in a sea of vague generalization; and all because this subject has seldom been examined in a spirit of scientific investigation.

CHAPTER 2

ON THE THEORY OF WAR

Originally the Term 'Art of War' Only Designated the Preparation of the Forces

FORMERLY, the terms 'art of war' or 'science of war'* were used to designate only the total body of knowledge and skill that was concerned with material factors. The design, production, and use of weapons, the construction of fortifications and entrenchments, the internal organization of the army, and the mechanism of its movements constituted the substance of this knowledge and skill. All contributed to the establishment of an effective fighting force. It was a case of handling a material substance, a unilateral activity, and was basically nothing but a gradual rise from a craft to a refined mechanical art. It was about as relevant to combat as the craft of the swordsmith to the art of fencing. It did not yet include the use of force under conditions of danger, subject to constant interaction with an adversary, nor the efforts of spirit and courage to achieve a desired end.

True War First Appears in Siege Warfare

Siege warfare gave the first glimpse of the conduct of operations, of intellectual effort; but this usually revealed itself only in such new techniques as approaches, trenches, counterapproaches, batteries and so forth, and marked each step by some such product. It was only the thread needed to link these material inventions. Since in siege warfare that is almost the only way in which the intellect can manifest itself, the matter usually rested there.

Next the Subject Was Touched on by Tactics

Later, tactics attempted to convert the structure of its component parts into a general system, based on the peculiar properties of its instrument.* This certainly led to the battlefield, but not yet to creative intellectual activity. The result was rather armies which had been transformed by their formations and orders of battle into automata, designed to discharge their activity like pieces of clockwork set off by a mere word of command.

The Actual Conduct of War Occurred Only Incidentally and Incognito

The actual conduct of war—the free use of the given means, appropriate to each individual occasion—was not considered a suitable subject for theory, but one that had to be left to natural preference. Gradually, war progressed from medieval hand-to-hand fighting toward a more orderly and complex form. Then, admittedly, the human mind was forced to give some thought to this matter; but as a rule its reflections appear only incidentally and, so to speak, incognito, in memoirs and histories.

Reflections on the Events of War Led to the Need for a Theory

As these reflections grew more numerous and history more sophisticated, an urgent need arose for principles and rules whereby the controversies that are so normal in military history—the debate between conflicting opinions—could be brought to some sort of resolution. This maelstrom of opinions, lacking in basic principles and clear laws round which they could be crystallized, was bound to be intellectually repugnant.

Efforts To Formulate a Positive Theory

Efforts were therefore made to equip the conduct of war with principles, rules, or even systems. This did present a positive goal, but people failed to take adequate account of the endless complexities involved. As we have seen, the conduct of war branches out in almost all directions and has no definite limits; while any system, any model, has the finite nature of a synthesis. An irreconcilable conflict exists between this type of theory and actual practice.

Limitation to Material Factors

Theorists soon found out how difficult the subject was, and felt justified in evading the problem by again directing their principles and systems only to physical matters and unilateral activity. As in the science concerning *preparation for war*, they wanted to reach a set of sure and positive conclusions, and for that reason considered only factors that could be mathematically calculated.

Numerical Superiority

Numerical superiority was a material factor. It was chosen from all elements that make up victory because, by using combinations of time and space, it could be fitted into a mathematical system of laws. It was thought that all other factors could be ignored if they were assumed to be equal on both sides and thus cancelled one another out. That might have been acceptable as a temporary device for the study of the characteristics of this single factor; but to make the device permanent, to accept superiority of numbers as the one and only rule, and to reduce the whole secret of the art of war to the formula of numerical superiority *at a certain time in a certain place* was an oversimplification that would not have stood up for a moment against the realities of life.

Supply

Another theoretical treatment sought to reduce a different material factor to a system: supply. Based on the assumption that an army was organized in a certain manner, its supply was set up as a final arbiter for the conduct of war.

That approach also produced some concrete figures, but these rested on a mass of arbitrary assumptions. They were therefore not able to stand the test of practical experience.

Base

One ingenious mind* sought to condense a whole array of factors, some of which did indeed stand in intellectual relation to one another, into a single concept, that of the *base*. This included *feeding the army, replacing its losses in men and equipment, assuring its communications with home, and even the safety of its retreat* in case that should become necessary. He started by substituting this concept for all these individual factors; next substituting the area or extent of this base for the concept itself, and ended up by substituting for this area the angle which the fighting forces created with their base line. All this led to a purely geometrical result, which is completely useless. This uselessness is actually inevitable in view of the fact that none of these substitutions could be made without doing violence to the facts and without dropping part of the content of the original idea. The concept of a base is a necessary tool in strategy and the author deserves credit for having discovered it; but it is completely inadmissible to use it in the manner described. It was bound to lead to one-sided conclusions which propelled that theorist into the rather contradictory direction of believing in the superior effectiveness of enveloping positions.

Interior Lines

As a reaction to that fallacy, another geometrical principle was then exalted:* that of so-called interior lines. Even though this tenet rests on solid ground—on the fact that the engagement is the only effective means in war—its purely geometrical character, still makes it another lopsided principle that could never govern a real situation.

All These Attempts Are Objectionable

It is only analytically that these attempts at theory can be called advances in the realm of truth; synthetically, in the rules and regulations they offer, they are absolutely useless.

They aim at fixed values; but in war everything is uncertain, and calculations have to be made with variable quantities.

They direct the inquiry exclusively toward physical quantities, whereas all military action is intertwined with psychological forces and effects.

They consider only unilateral action, whereas war consists of a continuous interaction of opposites.

They Exclude Genius from the Rule

Anything that could not be reached by the meagre wisdom of such one-sided points of view was held to be beyond scientific control: it lay in the realm of genius, *which rises above all rules*.

Pity the soldier who is supposed to crawl among these scraps of rules, not good enough for genius, which genius can ignore, or laugh at. No; what genius does is the best rule, and theory can do no better than show how and why this should be the case.

Pity the theory that conflicts with reason! No amount of humility can gloss over this contradiction; indeed, the greater the humility, the sooner it will be driven off the field of real life by ridicule and contempt.

Problems Facing Theory When Moral Factors Are Involved

Theory becomes infinitely more difficult as soon as it touches the realm of moral values. Architects and painters know precisely what they are about as long as they deal with material phenomena. Mechanical and optical structures are not subject to dispute. But when they come to the aesthetics of their work, when they aim at a particular effect on the mind or on the senses, the rules dissolve into nothing but vague ideas.

Medicine is usually concerned only with physical phenomena. It deals with the animal organism, which, however, is subject to constant change, and thus is never exactly the same from one moment to the next. This renders the task of medicine very difficult, and makes the physician's judgement count for more than his knowledge. But how greatly is the difficulty increased when a mental factor is added, and how much more highly do we value the psychiatrist!

Moral Values Cannot Be Ignored in War

Military activity is never directed against material force alone; it is always aimed simultaneously at the moral forces which give it life, and the two cannot be separated.

But moral values can only be perceived by the inner eye, which differs in each person, and is often different in the same person at different times.

Since danger is the common element in which everything moves in war, courage, the sense of one's own strength, is the principal factor that influences judgement. It is the lens, so to speak, through which impressions pass to the brain.

And yet there can be no doubt that experience will by itself provide a degree of objectivity to these impressions.

Everyone knows the moral effects of an ambush or an attack in flank or rear. Everyone rates the enemy's bravery lower once his back is turned, and takes much greater risks in pursuit than while being pursued. Everyone gauges his opponent in the light of his reputed talents, his age, and his experience, and acts accordingly. Everyone tries to assess the spirit and temper of his own troops and of the enemy's. All these and similar effects in the sphere of mind and spirit have been proved by experience: they recur constantly, and are therefore entitled to receive their due as objective factors. What indeed would become of a theory that ignored them?

Of course these truths must be rooted in experience. No theorist, and no commander, should bother himself with psychological and philosophical sophistries.

Principal Problems in Formulating a Theory of the Conduct of War

In order to get a clear idea of the difficulties involved in formulating a theory of the conduct of war and so be able to deduce its character, we must look more closely at the major characteristics of military activity.

First Property: Moral Forces and Effects

Hostile feelings

The first of these attributes consists of moral forces and the effects they produce.

Essentially combat is an expression of *hostile feelings*. But in the large-scale combat that we call war hostile feelings often have become merely hostile *intentions*. At any rate there are usually no hostile feelings between individuals. Yet such emotions can never be completely absent from war. Modern wars are seldom fought without hatred between nations; this serves more or less as a substitute for hatred between individuals. Even where there is no national hatred and no animosity to start with, the fighting itself will stir up hostile feelings: violence committed on superior orders will stir up the desire for revenge and retaliation against the perpetrator rather than against the powers that ordered the action. That is only human (or animal, if you like), but it is a fact. Theorists are apt to look on fighting in the abstract as a trial of strength without emotion entering into it. This is one of a thousand errors which they quite consciously commit because they have no idea of the implications.

Apart from emotions stimulated by the nature of combat, there are others that are not so intimately linked with fighting; but because of a certain affinity, they are easily associated with fighting: ambition, love of power, enthusiasms of all kinds, and so forth.

The Effects of Danger

Courage

Combat gives rise to the element of danger in which all military activity must move and be maintained like birds in air and fish in water. The effects of danger, however, produce an emotional reaction, either as a matter of immediate instinct, or consciously. The former results in an effort to avoid the danger, or, where that is not possible, in fear and anxiety. Where these effects do not arise, it is because instinct has been outweighed by *courage*. But courage is by no means a conscious act; like fear, it is an emotion. Fear is concerned with physical and courage with moral survival. Courage is the nobler instinct, and as such cannot be treated as an inanimate instrument that functions simply as prescribed. So courage is not simply a counterweight to danger, to be used for neutralizing its effects: it is a quality on its own.

Extent of the Influence Exercised by Danger

In order properly to appreciate the influence which danger exerts in war, one should not limit its sphere to the physical hazards of the moment. Danger dominates the commander not merely by threatening him personally, but by threatening all those entrusted to him; not only at the moment where it is actually present, but also, through the imagination, at all other times when it is relevant; not just directly but also indirectly through the sense of responsibility that lays a tenfold burden on the commander's mind. He could hardly recommend or decide on a major battle without a certain feeling of strain and distress at the thought of the danger and responsibility such a major decision implies. One can make the point that action in war, insofar as it is true action and not mere existence, is never completely free from danger.

Other Emotional Factors

In considering emotions that have been aroused by hostility and danger as being peculiar to war, we do not mean to exclude all others that accompany man throughout his life. There is a place for them in war as well. It may be true that many a petty play of emotions is silenced by the serious duties of war; but that holds only for men in the lower ranks who, rushed from one set of exertions and dangers to the next, lose sight of the other things in life, forgo duplicity because death will not respect it, and thus arrive at the soldierly simplicity of character that has always represented the military at its best. In the higher ranks it is different. The higher a man is placed, the broader his point of view. Different interests and a wide variety of passions, good and bad, will arise on all sides. Envy and generosity, pride and humility, wrath and compassion—all may appear as effective forces in this great drama.

Intellectual Qualities

In addition to his emotional qualities, the intellectual qualities of the commander are of major importance. One will expect a visionary, high-flown and immature mind to function differently from a cool and powerful one.

The Diversity of Intellectual Quality Results in a Diversity of Roads to the Goal

The influence of the great diversity of intellectual qualities is felt chiefly in the higher ranks, and increases as one goes up the ladder. It is the primary cause for the diversity of roads to the goal—already discussed in Book I—and for the disproportionate part assigned to the play of probability and chance in determining the course of events.

Second Property: Positive Reaction

The second attribute of military action is that it must expect positive reactions, and the process of interaction that results. Here we are not concerned with the problem of calculating such reactions—that is really part of the already mentioned problem of calculating psychological forces—but rather with the fact that the very nature of interaction is bound to make it unpredictable. The effect that any measure will have on the enemy is the most singular factor among all the particulars of action. All theories, however, must stick to categories of phenomena and can never take account of a truly unique case; this must be left to judgement and talent. Thus it is natural that military activity, whose plans, based on general circumstances, are so frequently disrupted by unexpected particular events, should remain largely a matter of talent, and that theoretical *directives* tend to be less useful here than in any other sphere.

Third Property: Uncertainty of All Information

Finally, the general unreliability of all information presents a special problem in war: all action takes place, so to speak, in a kind of twilight, which, like fog or moonlight, often tends to make things seem grotesque and larger than they really are.

Whatever is hidden from full view in this feeble light has to be guessed at by talent, or simply left to chance. So once again for lack of objective knowledge one has to trust to talent or to luck.

A Positive Doctrine Is Unattainable

Given the nature of the subject, we must remind ourselves that it is simply not possible to construct a model for the art of war that can serve as a scaffolding on which the commander can rely for support at any time. Whenever he has to fall back on his innate talent, he will find himself outside the model and in conflict with it; no matter how versatile the code, the situation will always lead to the consequences we have already alluded to: *talent and genius operate outside the rules, and theory conflicts with practice.*

Alternatives Which Make a Theory Possible

The difficulties vary in magnitude

There are two ways out of this dilemma.

In the first place, our comments on the nature of military activity in general should not be taken as applying equally to action at all levels. What is most needed in the lower ranks is courage and self-sacrifice, but there are far fewer problems to be solved by intelligence and judgement. The field of action is more limited, means and ends are fewer in number, and the data more concrete: usually they are limited to what is actually visible. But the higher the rank, the more the problems multiply, reaching their highest point in the supreme commander. At this level, almost all solutions must be left to imaginative intellect.

Even if we break down war into its various *activities*, we will find that the difficulties are not uniform throughout. The more physical the activity, the less the difficulties will be. The more the activity becomes intellectual and turns into motives which exercise a determining influence on the commander's will, the more the difficulties will increase. Thus it is easier to use theory to organize, plan, and conduct an engagement than it is to use it in determining the engagement's purpose. Combat is conducted with physical weapons, and although the intellect does play a part, material factors will dominate. But when one comes to the *effect* of the engagement, where material successes turn into motives for further action, the intellect alone is decisive. In brief, *tactics* will present far fewer difficulties to the theorist than will *strategy*.

Theory Should Be Study, Not Doctrine

The second way out of this difficulty is to argue that a theory need not be a positive doctrine, a sort of *manual* for action. Whenever an activity deals primarily with the same things again and again—with the same ends and the same means, even though there may be minor variations and an infinite diversity of combinations—these things are susceptible of rational study. It is precisely that inquiry which is the most essential part of any *theory*, and which may quite appropriately claim that title. It is an analytical investigation leading to a close *acquaintance* with the subject; applied to experience—in our case, to military history—it leads to thorough *familiarity* with it. The closer it comes to that goal, the more it proceeds from the objective form of a science to the subjective form of a skill, the more effective it will prove in areas where the nature of the case admits no arbiter but talent. It will, in fact, become an active ingredient of talent. Theory will have fulfilled its main task when it is used to analyse the constituent elements of war, to distinguish precisely what at first sight seems fused, to explain in full the properties of the means employed and to show their probable effects, to define clearly the nature of the ends in view, and to illuminate all phases of warfare in a thorough critical inquiry. Theory then becomes a guide to anyone who wants to learn about war from books; it will light his way, ease his progress, train his judgement, and help him to avoid pitfalls.

A specialist who has spent half his life trying to master every aspect of some obscure subject is surely more likely to make headway than a man who is trying to master it in a short time. Theory exists so that one need not start afresh each time sorting out the material and ploughing through it, but will find it ready to hand and in good order. It is meant to educate the mind of the future commander, or, more accurately, to guide him in his self-education, not to accompany him to the battlefield; just as a wise teacher guides and stimulates a young man's intellectual development, but is careful not to lead him by the hand for the rest of his life.

If the theorist's studies automatically result in principles and rules, and if truth spontaneously crystallizes into these forms, theory will not resist this natural tendency of the mind. On the contrary, where the arch of truth culminates in such a keystone, this tendency will be underlined. But this is simply in accordance with the scientific law of reason, to indicate the point at which all lines converge, but never to construct an algebraic formula for use on the battlefield. Even these principles and rules are intended to provide a thinking man with a frame of reference for the movements he has been trained to carry out, rather than to serve as a guide which at the moment of action lays down precisely the path he must take.

This Point of View Makes Theory Possible and Eliminates Its Conflict with Reality

This point of view will admit the feasibility of a satisfactory theory of war—one that will be of real service and will never conflict with reality. It only needs intelligent treatment to make it conform to action, and to end the absurd difference between theory and practice that unreasonable theories have so often evoked. That difference, which defies common sense, has often been used as a pretext by limited and ignorant minds to justify their congenital incompetence.

Theory Thus Studies The Nature of Ends and Means

Ends and means in tactics

It is the task of theory, then, to study the nature of ends and means.

In tactics the means are the fighting forces trained for combat; the end is victory. A more precise definition of this concept will be offered later on, in the context of 'the engagement'. Here, it is enough to say that the enemy's withdrawal from the battlefield is the sign of victory. Strategy thereby gains the end it had ascribed to the engagement, the end that constitutes its real *significance*. This significance admittedly will exert a certain influence on the kind of victory achieved. A victory aimed at weakening the enemy's fighting forces is different from one that is only meant to seize a certain position. The significance of an engagement may therefore have a noticeable influence on its planning and conduct, and is therefore to be studied in connection with tactics.

Factors That Always Accompany the Application of the Means

There are certain constant factors in any engagement that will affect it to some extent; we must allow for them in our use of armed forces.

These factors are the locality or terrain, the time of day, and the weather.

Terrain

Terrain, which can be resolved into a combination of the geographical surroundings and the nature of the ground, could, strictly speaking, be of no influence at all on an engagement fought over a flat, uncultivated plain.

This does actually occur in the steppes, but in the cultivated parts of Europe it requires an effort of the imagination to conceive it. Among civilized nations combat uninfluenced by its surroundings and the nature of the ground is hardly conceivable.

Time of Day

The time of day affects an engagement by the difference between day and night. By implication, of course, these precise limits may be exceeded: every engagement takes a certain time, and major ones may last many hours. When a major battle is being planned, it makes a decisive difference whether it is to start in the morning or in the afternoon. On the other hand there are many engagements where the time of day is a neutral factor; in the general run of cases it is of minor importance.

Weather

It is rarer still for weather to be a decisive factor. As a rule only fog makes any difference.

Ends and Means in Strategy

The original means of strategy is victory—that is, tactical success; its ends, in the final analysis, are those objects which will lead directly to peace. The application of these means for these ends will also be attended by factors that will influence it to a greater or lesser degree.

Factors That Affect the Application of the Means

These factors are the geographical surroundings and nature of the terrain (the former extended to include the country and people of the entire theatre of war); the time of day (including the time of year); and the weather (particularly unusual occurrences such as severe frost, and so forth).

These Factors Form New Means

Strategy, in connecting these factors with the outcome of an engagement, confers a special significance on that outcome and thereby on the engagement: *it assigns a particular aim to it*. Yet insofar as that aim is not the one that will lead directly to peace, it remains subsidiary and is also to be thought of as a means. Successful engagements or victories in all stages of importance may therefore be considered as strategic means. The capture of a position is a successful engagement in terms of terrain. Not only individual engagements with particular aims are to be classified as means: any greater unity formed in a combination of engagements by being directed toward a common aim can also be considered as *a means*. A winter campaign is such a combination in terms of the time of year.

What remains in the way of ends, then, are only those objects that lead *directly* to peace. All these ends and means must be examined by the theorist in accordance with their effects and their relationships to one another.

Strategy Derives the Means and Ends To Be Examined Exclusively from Experience

The first question is, how an exhaustive list of these objects is arrived at. If a scientific examination were meant to produce this result, it

would become involved in all those difficulties which logical necessity has excluded both from the conduct and from the theory of war.
We therefore turn to experience and study those sequences of events
related in military history. The result will, of course, be a limited
theory, based only on facts recorded by military historians. But that
is inevitable, since theoretical results must have been derived from
military history or at least checked against it. Such a limitation is in
any case more theoretical than real.

A great advantage offered by this method is that theory will
have to remain realistic. It cannot allow itself to get lost in futile
speculation, hairsplitting, and flights of fancy.

How Far Should an Analysis of the Means be Carried?

A second question is, how far theory should carry its analysis of the
means. Obviously only so far as the separate attributes will have
significance in practice. The range and effectiveness of different firearms is tactically most important; but their construction, though it
governs their performance, is irrelevant. The conduct of war has
nothing to do with making guns and powder out of coal, sulphur,
saltpetre, copper and tin; its given quantities are weapons that are
ready for use and their effectiveness. Strategy uses maps without
worrying about trigonometric surveys; it does not inquire how a
country should be organized and a people trained and ruled in order
to produce the best military results. It takes these matters as it finds
them in the European community of nations, and calls attention only
to unusual circumstances that exert a marked influence on war.

Substantial Simplification of Knowledge

Clearly, then, the range of subjects a theory must cover may be
greatly simplified and the knowledge required for the conduct of war
can be greatly reduced. Military activity in general is served by an
enormous amount of expertise and skills, all of which are needed to
place a well-equipped force in the field. They coalesce into a few
great results before they attain their final purpose in war, like
streams combining to form rivers before they flow into the sea. The
man who wishes to control them must familiarize himself only with
those activities that empty themselves into the great ocean of war.

This Simplification Explains the Rapid Development of Great Commanders, and Why Commanders Are Not Scholars

In fact, this result of our investigation is so inescapable that if it were any different its validity would be in doubt. Only this explains why in war men have so often successfully emerged in the higher ranks, and even as supreme commanders, whose former field of endeavour was entirely different; the fact, indeed, that distinguished commanders have never emerged from the ranks of the most erudite or scholarly officers, but have been for the most part men whose station in life could not have brought them a high degree of education. That is why anyone who thought it necessary or even useful to begin the education of a future general with a knowledge of all the details has always been scoffed at as a ridiculous pedant. Indeed, that method can easily be proved to be harmful: for the mind is formed by the knowledge and the direction of ideas it receives and the guidance it is given. Great things alone can make a great mind, and petty things will make a petty mind unless a man rejects them as completely alien.

Earlier contradictions

The simplicity of the knowledge required in war has been ignored: or rather, that knowledge has always been lumped together with the whole array of ancillary information and skills. This led to an obvious contradiction with reality, which could only be resolved by ascribing everything to genius that needs no theory and for which no theory ought to be formulated.

Accordingly, the usefulness of all knowledge was denied, and everything was ascribed to natural aptitude

Everyone with a grain of common sense realized the vast distance between a genius of the highest order and a learned pedant. Men arrived at a type of free thinking that rejected all belief in theory and postulated that the conduct of war was a natural function of man which he performed as well as his aptitude permitted. It cannot be denied that this view is closer to the truth than the emphasis on irrelevant expertise; still, on closer examination it will be found to be an overstatement. No activity of the human mind is possible without a certain stock of ideas; for the most part these are not innate but acquired, and constitute a man's knowledge. The only question therefore is what type of ideas they should be. We believe that we have answered this by saying that they should refer only to those things with which he will be immediately concerned as a soldier.

Knowledge Will Be Determined by Responsibility

Within this field of military activity, ideas will differ in accordance with the commander's area of responsibility. In the lower ranks they will be focused upon minor and more limited objectives; in the more senior, upon wider and more comprehensive ones. There are commanders-in-chief who could not have led a cavalry regiment with distinction, and cavalry commanders who could not have led armies.

The Knowledge Required in War Is Very Simple, but at the Same Time It Is Not Easy to Apply

Knowledge in war *is very simple*, being concerned with so few subjects, and only with their final results at that. But this does not make its application easy. The obstacles to action in general have already been discussed in Book I. Leaving aside those that can be overcome only by courage, we argue that genuine intellectual activity is simple and easy only in the lower ranks. The difficulty increases with every step up the ladder; and at the top—the position of commander-in-chief—it becomes among the most extreme to which the mind can be subjected.

The Nature of Such Knowledge

A commander-in-chief need not be a learned historian nor a pundit, but he must be familiar with the higher affairs of state and its innate policies; he must know current issues, questions under consideration, the leading personalities, and be able to form sound judgements. He need not be an acute observer of mankind or a subtle analyst of human character; but he must know the character, the habits of thought and action, and the special virtues and defects of the men whom he is to command. He need not know how to manage a wagon or harness a battery horse, but he must be able to gauge how long a column will take to march a given distance under various conditions. This type of knowledge cannot be forcibly produced by an apparatus of scientific formulas and mechanics; it can only be gained through a talent for judgement, and by the application of accurate judgement to the observation of man and matter.

The knowledge needed by a senior commander is distinguished by the fact that it can only be attained by a special talent, through the medium of reflection, study and thought: an intellectual instinct which extracts the essence from the phenomena of life, as a bee sucks honey from a flower. In addition to study and reflection, life itself serves as a source. Experience, with its wealth of lessons, will never produce a *Newton* or an *Euler*,* but it may well bring forth the higher calculations of a *Condé* or a *Frederick*.*

To save the intellectual repute of military activity there is no need to resort to lies and simple-minded pedantry. No great commander was ever a man of limited intellect. But there are numerous cases of men who served with the greatest distinction in the lower ranks and turned out barely mediocre in the highest commands, because their intellectual powers were inadequate. Even among commanders-in-chief a distinction must of course be made according to the extent of their authority.

Knowledge Must Become Capability

One more requisite remains to be considered—a factor more vital to military knowledge than to any other. Knowledge must be so absorbed into the mind that it almost ceases to exist in a separate, objective way. In almost any other art or profession a man can work with truths he has learned from musty books, but which have no life or meaning for him. Even truths that are in constant use and are always to hand may still be externals. When an architect sits down with pen and paper to determine the strength of an abutment by a complicated calculation, the truth of the answer at which he arrives is not an expression of his own personality. First he selects the data with care, then he submits them to a mental process not of his own invention, of whose logic he is not at the moment fully conscious, but which he applies for the most part mechanically. It is never like that in war. Continual change and the need to respond to it compels the commander to carry the whole intellectual apparatus of his knowledge within him. He must always be ready to bring forth the appropriate decision. By total assimilation with his mind and life, the commander's knowledge must be transformed into a genuine capability. That is why it all seems to come so easily to men who have distinguished themselves in war, and why it is all ascribed to natural talent. We say *natural talent* in order to distinguish it from the talent that has been trained and educated by reflection and study.

These observations have, we believe, clarified the problems that confront any theory of warfare, and suggested an approach to its solution.

We have divided the conduct of war into the two fields of tactics and strategy. The theory of the latter, as we have already stated, will unquestionably encounter the greater problems since the former is virtually limited to material factors, whereas for strategic theory, dealing as it does with ends which bear directly on the restoration of peace, the range of possibilities is unlimited. As these ends will have to be considered primarily by the *commander-in-chief*, the problems mainly arise in those fields that lie within his competence.

In the field of strategy, therefore, even more than in tactics, theory will be content with the simple consideration of material and psychological factors, especially where it embraces the highest of achievements. It will be sufficient if it helps the commander acquire those insights that, once absorbed into his way of thinking, will smooth and protect his progress, and will never force him to abandon his convictions for the sake of any objective fact.

CHAPTER 3

ART OF WAR OR SCIENCE OF WAR*

Usage Is Still Unsettled

Ability and knowledge. The object of science is knowledge; the object of art is creative ability

THE use of these terms seems still to be unsettled, and simple though the matter may be, we apparently still do not know on what basis we should choose between them. We have already argued that *knowledge* and *ability* are different things—so different that there should be no cause for confusion. A book cannot really teach us how to do anything, and therefore 'art' should have no place in its title. But we have become used to summarizing the knowledge required for the practice of art (individual branches of which may be complete sciences in themselves) by the term 'theory of art', or simply 'art'. It is therefore consistent to keep this basis of distinction and call everything 'art' whose object is creative ability, as, for instance, architecture. The term 'science' should be kept for disciplines such as mathematics or astronomy, whose object is pure knowledge. That every theory of art may contain discrete sciences goes without saying, and need not worry us. But it is also to be noted that no science can exist without some element of art: in mathematics, for instance, the use of arithmetic and algebra is an art. But art may go still further. The reason is that, no matter how obvious and palpable the difference between knowledge and ability may be in the totality of human achievement, it is still extremely difficult to separate them entirely in the individual.

The Difficulty of Separating Perception from Judgement

Art of war

Of course all thought is art. The point where the logician draws the line, where the premises resulting from perceptions end and where judgement starts, is the point where art begins. But further: perception by the mind is already a judgement and therefore an art; so too, in the last analysis, is perception by the senses. In brief, if it is impossible to imagine a human being capable of perception but not of judgement or vice versa, it is likewise impossible to separate art and knowledge altogether. The more these delicate motes of light are personified in *external forms* of being, the more will their realms separate. To repeat, creation and production lie in the realm of art; science will dominate where the object is inquiry and knowledge. It follows that the term 'art of war' is more suitable than 'science of war'.

We have discussed this at length because these concepts are indispensable. But we must go on to say that strictly speaking war is neither an art nor a science. To take these concepts as a point of departure is misleading in that it has unintentionally caused war to be put on a par with other arts or sciences, resulting in a mass of incorrect analogies.

This difficulty was already recognized in the past, and it was therefore suggested that war was a craft. That, however, proved more of a loss than a gain, because a craft is simply an *inferior* form of art and as such subject to stricter and more rigorous laws. Actually, there was a time—the age of the *condottieri**—when the art of war was akin to a craft. But this tendency had no *internal*, only an *external* basis. Military history shows how unnatural and unsatisfactory it turned out to be.

War Is an Act of Human Intercourse

We therefore conclude that war does not belong in the realm of arts and sciences; rather it is part of man's social existence. War is a clash between major interests, which is resolved by bloodshed—that is the only way in which it differs from other conflicts. Rather than comparing it to art we could more accurately compare it to commerce, which is also a conflict of human interests and activities; and it is *still* closer to politics, which in turn may be considered as a kind of commerce on a larger scale. Politics, moreover, is the womb in which war develops—where its outlines already exist in their hidden rudimentary form, like the characteristics of living creatures in their embryos.

Difference

The essential difference is that war is not an exercise of the will directed at inanimate matter, as is the case with the mechanical arts, or at matter which is animate but passive and yielding, as is the case with the human mind and emotions in the fine arts. In war, the will is directed at an animate object that *reacts*. It must be obvious that the intellectual codification used in the arts and sciences is inappropriate to such an activity. At the same time it is clear that continual striving after laws analogous to those appropriate to the realm of inanimate matter was bound to lead to one mistake after another. Yet it was precisely the mechanical arts that the art of war was supposed to imitate. The fine arts were impossible to imitate, since they themselves do not yet have sufficient laws and rules of their own. So far all attempts at formulating any have been found too limited and one-sided and have constantly been undermined and swept away by the currents of opinion, emotion and custom.

Part of the object of this book is to examine whether a conflict of living forces as it develops and is resolved in war remains subject to general laws, and whether these can provide a useful guide to action. This much is clear: this subject, like any other that does not surpass man's intellectual capacity, can be elucidated by an inquiring mind, and its internal structure can to some degree be revealed. That alone is enough to turn the concept of theory into reality.

CHAPTER 4

METHOD AND ROUTINE

IN order to explain succinctly the concepts of method and routine, which play such an important role in war, we must glance briefly at the logical hierarchy that governs the world of action like a duly constituted authority.

Law is the broadest concept applicable to both perception and action. In its literal sense the term obviously contains a subjective, arbitrary element, and yet it expresses the very thing on which man and his environment essentially depend. Viewed as a matter of cognition, law is the relationship between things and their effects. Viewed as a matter of the will, law is a determinant of action; at that point, it is synonymous with *decree* and *prohibition*.

Principle is also a law for action, but not in its *formal, definitive meaning*; it represents only the spirit and the sense of the law: in cases where the diversity of the real world cannot be contained within the rigid form of law, the application of principle allows for a greater latitude of judgement. Cases to which principle cannot be applied must be settled by judgement; principle thus becomes essentially a support, or lodestar, to the man responsible for the action.

A principle is *objective* if it rests on objective truth and is therefore equally valid for all; it is *subjective* and is generally called a maxim if subjective considerations enter into it. In that case it has value only for the person who adopts it.

Rule is a term often used in the sense of law; it then becomes synonymous with principle. The proverb goes 'there is an exception to every rule' and not 'to every law', which shows that in the case of a rule one reserves the right to a more liberal interpretation.

In another sense, the term 'rule' is used for 'means': to recognize an underlying truth through a single obviously relevant feature enables us to derive a general law of action from this feature. Rules in games are like this, and so are the short cuts used in mathematics, and so on.

Regulations and *directions* are directives dealing with a mass of minor, more detailed circumstances, too numerous and too trivial for general laws.

'*Method*', finally, or 'mode of procedure', is a constantly recurring procedure that has been selected from several possibilities. It becomes routine when action is prescribed by method rather than by general principles or individual regulation. It must necessarily be assumed that all cases to which such a routine is applied will be essentially alike. Since this will not be entirely so, it is important that it be true of at least *as many as possible*. In other words, methodical procedure should be designed to meet the most probable cases. Routine is not based on definite individual premises, but rather on the *average probability* of analogous cases. Its aim is to postulate an average truth, which, when applied evenly and constantly, will soon acquire some of the nature of a mechanical skill, which eventually does the right thing almost automatically.

In the conduct of war, perception cannot be governed by laws: the complex phenomena of war are not so uniform, nor the uniform phenomena so complex, as to make laws more useful than the simple truth. Where a simple point of view and plain language are sufficient, it would be pedantic and affected to make them *complex* and *involved*. Nor can the theory of war apply the concept of law to action, since no prescriptive formulation universal enough to deserve the name of law can be applied to the constant change and diversity of the phenomena of war.

Principles, rules, regulations, and methods are, however, indispensable concepts to or for that part of the theory of war that leads to positive doctrines; for in these doctrines the truth can express itself only in such compressed forms.

Those concepts will appear most frequently in tactics, which is that part of war in which theory can develop most fully into a positive doctrine. Some examples of tactical principles are: except in emergencies cavalry is not to be used against unbroken infantry; firearms should not be used until the enemy is within effective range; in combat, as many troops as possible should be preserved for the final phase. None of these concepts can be dogmatically applied to every situation, but a commander must always bear them in mind so as not to lose the benefit of the truth they contain in cases where they do apply.

Cooking in the enemy camp at unusual times suggests that he is about to move. The intentional exposure of troops in combat indicates a feint. This manner of inferring the truth may be called a rule because one deduces the enemy's intentions from a single visible fact connected with them.

If the rule enjoins that one should resume attacking the enemy as soon as he starts to withdraw his artillery, then a whole course of action is determined by this single phenomenon which has revealed his entire condition: the fact that he is ready to give up the fight. While he is doing so, he cannot offer serious resistance or even avoid action as he could once he is fully on the move.

To the extent that *regulations* and *methods* have been drilled into troops as active principles, theoretical preparation for war is part of its actual conduct. All standing instructions on formations, drill, and field-service are regulations and methods. Drill instructions are mainly regulations; field manuals, mainly methods. The actual conduct of war is based on these things; they are accepted as given procedures and as such must have their place in the theory of the conduct of war.

In the employment of these forces, some activities remain a matter of choice. Regulations, or prescriptive directions, do not apply to them, precisely because regulations preclude freedom of choice. Routines, on the other hand, represent a general way of executing tasks as they arise based, as we have said, on average probability. They represent the dominance of principles and rules, carried through to actual application. As such they may well have a place in the theory of the conduct of war, provided they are not falsely represented as absolute, binding frameworks for action (systems); rather they are the best of the general forms, short cuts, and options that may be substituted for individual decisions.

The frequent application of routine in war will also appear essential and inevitable when we consider how often action is based on

pure conjecture or takes place in complete ignorance, either because
the enemy prevents us from knowing all the circumstances that
might affect our dispositions, or because there is not enough time.
Even if we did know all the circumstances, their implications and
complexities would not permit us to take the necessary steps to deal
with them. Therefore our measures must always be determined by a
limited number of possibilities. We have to remember the countless
minor factors implicit in every case. The only possible way of dealing
with them is to treat each case as implying all the others, and base
our dispositions on the general and the probable. Finally we have to
remember that as the number of officers increases steadily in the
lower ranks, the less the trust that can be placed on their true insight
and mature judgement. Officers whom one should not expect to have
any greater understanding than regulations and experience can give
them have to be helped along by routine methods tantamount to
rules. These will steady their judgement, and also guard them
against eccentric and mistaken schemes, which are the greatest
menace in a field where experience is so dearly bought.

Routine, apart from its sheer inevitability, also contains one positive advantage. Constant practice leads to *brisk, precise*, and *reliable*
leadership, reducing natural friction and easing the working of
the machine.

In short, routine will be more frequent and indispensable, the
lower the level of action. As the level rises, its use will decrease to the
point where, at the summit, it disappears completely. Consequently,
it is more appropriate to tactics than to strategy.

War, in its highest forms, is not *an infinite mass of minor events*,
analogous despite their diversities, which can be controlled with
greater or lesser effectiveness depending on the methods applied.
War consists rather of *single, great decisive actions*, each of which
needs to be handled individually. War is not like a field of wheat,
which, without regard to the individual stalk, may be mown more or
less efficiently depending on the quality of the scythe; it is like a
stand of mature trees in which the axe has to be used judiciously
according to the characteristics and development of each individual
trunk.

The highest level that routine may reach in military action is of
course determined not by rank but by the nature of each situation.
The highest ranks are least affected by it simply because the scope of

their operations is the most comprehensive. A standard order of battle or system of advance guards and outposts are methods by which a general may be fettering not only his subordinates, but, in certain cases, also himself. Of course these methods may be his own inventions, and adapted to particular conditions; to the extent that they are based on the general properties of troops and weapons, they can also be a subject of theory. But any method by which strategic plans are turned out ready-made, as if from some machine, must be totally rejected.

So long as no acceptable theory, no intelligent analysis of the conduct of war exists, routine methods will tend to take over even at the highest levels. Some of the men in command have not had the opportunities of self-improvement afforded by education and contact with the higher levels of society and government. They cannot cope with the impractical and contradictory arguments of theorists and critics even though their common sense rejects them. Their only insights are those that have been gained by experience. For this reason, they prefer to use the means with which their experience has equipped them, even in cases that could and should be handled freely and individually. They will copy their supreme commander's favourite device—thus automatically creating a new routine. When we find generals under Frederick the Great using the so-called oblique order of battle;* generals of the French Revolution using turning movements with a much extended front; and commanders under Bonaparte attacking with a brutal rush of concentric masses, then we recognize in these repetitions a ready-made method, and see that even the highest ranks are not above the influence of routine. Once an improved theory helps the study of the conduct of war, and educates the mind and judgement of the senior commanders, routine methods will no longer reach so high. Those types of routine that must be considered indispensable will then at least be based on a theory rather than consist in sheer imitation. No matter how superbly a great commander operates, there is always a subjective element in his work. If he displays a certain style, it will in large part reflect his own personality; but that will not always blend with the personality of the man who copies that style.

Yet it would be neither possible nor correct to eliminate subjective routine or personal style entirely from the conduct of war. They should be seen, rather, as manifestations of the influence exerted on individual phenomena by the total character of the war—an influence which, if it has not been foreseen and allowed for by accepted theory, may find no other means of adequate expression. What could be more natural than the fact that the War of the French Revolution had its characteristic style, and what theory could have been expected to accommodate it? The danger is that this kind of style, developed out of a single case, can easily outlive the situation that gave rise to it; for conditions change imperceptibly. That danger is the very thing a theory should prevent by lucid, rational criticism. When in 1806 the Prussian generals, Prince Louis at Saalfeld, Tauentzien on the Dornberg near Jena, Grawert on one side of Kapellendorf and Rüchel* on the other, plunged into the open jaws of disaster by using Frederick the Great's oblique order of battle, it was not just a case of a style that had outlived its usefulness but the most extreme poverty of the imagination to which routine has ever led. The result was that the Prussian army under Hohenlohe* was ruined more completely than any army has ever been ruined on the battlefield.

CHAPTER 5

CRITICAL ANALYSIS

THE influence of theoretical truths on practical life is always exerted more through critical analysis than through doctrine. Critical analysis being the application of theoretical truths to actual events, it not only reduces the gap between the two but also accustoms the mind to these truths through their repeated application. We have established a criterion for theory, and must now establish one for critical analysis as well.

We distinguish between the *critical approach* and the plain narrative of a historical event, which merely arranges facts one after another, and at most touches on their immediate causal links.

Three different intellectual activities may be contained in the critical approach.

First, the discovery and interpretation of equivocal facts. This is historical research proper, and has nothing in common with theory.

Second, the tracing of effects back to their causes. This is *critical analysis proper.* It is essential for theory; for whatever in theory is to be defined, supported, or simply described by reference to experience can only be dealt with in this manner.

Third, the investigation and evaluation of means employed. This last is criticism proper, involving praise and censure. Here theory serves history, or rather the lessons to be drawn from history.

In the last two activities which are the truly critical parts of historical inquiry, it is vital to analyse everything down to its basic elements, to incontrovertible truth. One must not stop half-way, as is so often done, at some arbitrary assumption or hypothesis.

The deduction of effect from cause is often blocked by some insuperable extrinsic obstacle: the true causes may be quite unknown. Nowhere in life is this so common as in war, where the facts are seldom fully known and the underlying motives even less so. They may be intentionally concealed by those in command, or, if they happen to be transitory and accidental, history may not have recorded them at all. That is why critical narrative must usually go hand in hand with historical research. Even so, the disparity between cause and effect may be such that the critic is not justified in considering the effects as inevitable results of known causes. This is bound to produce gaps—historical results that yield no useful lesson. All a theory demands is that investigation should be resolutely carried on till such a gap is reached. At that point, judgement has to be suspended. Serious trouble arises only when known facts are forcibly stretched to explain effects; for this confers on these facts a spurious importance.

Apart from that problem, critical research is faced with a serious intrinsic one: effects in war seldom result from a single cause; there are usually several concurrent causes. It is therefore not enough to trace, however honestly and objectively, a sequence of events back to their origin: each identifiable cause still has to be correctly assessed. This leads to a closer analysis of the nature of these causes, and in this way critical investigation gets us into theory proper.

A critical *inquiry*—the examination of the means—poses the question as to what are the peculiar effects of the means employed, and whether these effects conform to the intention with which they were used.

The particular effects of the means leads us to an investigation of their nature—in other words, into the realm of theory again.

We have seen that in criticism it is vital to reach the point of incontrovertible truth; we must never stop at an arbitrary assumption that others may not accept, lest different propositions, equally valid perhaps, be advanced against them; leading to an unending argument, reaching no conclusions, and resulting in no lesson.

We have also seen that both investigation of the causes and examination of the means leads to the realm of theory—that is, to the field of universal truth that cannot be inferred merely from the individual instance under study. If a usable theory does indeed exist, the inquiry can refer to its conclusions and at that point end the investigation. However, where such theoretical criteria do not exist, analysis must be pressed until the basic elements are reached. If this happens often, it will lead the writer into a labyrinth of detail: he will have his hands full and find it almost impossible to give each point the attention it demands. As a result, in order to set a limit to his inquiries, he will have to stop short of arbitrary assumptions after all. Even if they would not seem arbitrary to him, they would to others, because they are neither self-evident nor have they been proved.

In short a working theory is an essential basis for criticism. Without such a theory it is generally impossible for criticism to reach that point at which it becomes truly instructive—when its arguments are convincing and cannot be refuted.

But it would be wishful thinking to imagine that any theory could cover every abstract truth, so that all the critic had to do would be to classify the case studied under the appropriate heading. Equally, it would be ridiculous to expect criticism to reverse course whenever it came up against the limits of a sacrosanct theory. The same spirit of analytical investigation which creates a theory should also guide the work of the critic who both may and should often cross into the realm of theory in order to elucidate any points of special importance. The function of criticism would be missed entirely if criticism were to degenerate into a mechanical application of theory. All the positive results of theoretical investigation—all the principles, rules, and methods—will increasingly lack universality and absolute truth the closer they come to being positive doctrine. They are there to be used when needed, and their suitability in any given case must always be a matter of judgement. A critic should never use the results of theory as laws and standards, but only—as the soldier does—as *aids to judgement*. If, in tactics, it is generally agreed that in the standard line of battle cavalry should be posted not in line with but behind the infantry, it would nevertheless be foolish to condemn every different deployment simply because it is different. The critic should analyse the reasons for the exception. He has no right to appeal to theoretical principles unless these reasons are inadequate. Again, if theory lays it down that an attack with divided forces reduces the probability of success, it would be equally unreasonable, without further analysis, to attribute failure to the separation of forces whenever both occur together; or when an attack with divided forces is successful to conclude that the original theoretical assertion was incorrect. The inquiring nature of criticism can permit neither. In short, criticism largely depends on the results of the theorist's analytic studies. What theory has already established the critic need not go over again, and it is the theorist's function to provide the critic with these findings.

The critic's task of investigating the relation of cause and effect and the appropriateness of means to ends will be easy when cause and effect, means and ends, are closely linked.

When a surprise attack renders an army incapable of employing its powers in an orderly and rational manner, then the effect of the surprise cannot be questioned. When theory has established that an enveloping attack leads to greater, if less certain, success, we have to ask whether the general who used this envelopment was primarily concerned with the magnitude of success. If so, he chose the right way to go about it. But if he used it in order to make *more certain* of success, basing his action not so much on individual circumstances as on the general nature of enveloping attacks, as has happened innumerable times, then he misunderstood the nature of the means he chose and committed an error.

The business of critical analysis and proof is not very difficult in cases of this kind; it is bound to be easy if one restricts oneself to the most immediate aims and effects. This may be done quite arbitrarily if one isolates the matter from its setting and studies it only under those conditions.

But in war, as in life generally, all parts of a whole are interconnected and thus the effects produced, however small their cause, must influence all subsequent military operations and modify their final outcome to some degree, however slight. In the same way, every means must influence even the ultimate purpose.

One can go on tracing the effects that a cause produces so long as it seems worth while. In the same way, a means may be evaluated, not merely with respect to its immediate end: that end itself should be appraised as a means for the next and highest one; and thus we can follow a chain of sequential objectives until we reach one that requires no justification, because its necessity is self-evident. In many cases, particularly those involving great and decisive actions, the analysis must extend to the *ultimate objective*, which is to bring about peace.

Every stage in this progression obviously implies a new basis for judgement. That which seems correct when looked at from one level may, when viewed from a higher one, appear objectionable.

In a critical analysis of the action, the search for the causes of phenomena and the testing of means in relation to ends always go hand in hand, for only the search for a cause will reveal the questions that need to be studied.

The pursuit of this chain, upward and downward, presents considerable problems. The greater the distance between the event and the cause that we are seeking, the larger the number of other causes that have to be considered at the same time. Their possible influence on events has to be established and allowed for, since the greater the magnitude of any event, the wider the range of forces and circumstances that affect it. When the causes for the loss of a battle have been ascertained, we shall admittedly also know some of the causes of the effects that this lost battle had upon the whole—but only some, since the final outcome will have been affected by other causes as well.

In the analysis of the means, we encounter the same multiplicity as our viewpoint becomes more comprehensive. The higher the ends, the greater the number of means by which they may be reached. The final aim of the war is pursued by all armies simultaneously, and we therefore have to consider the full extent of everything that has happened, or might have happened.

We can see that this may sometimes lead to a broad and complex field of inquiry in which we may easily get lost. A great many assumptions have to be made about things that did not actually happen but seemed possible, and that, therefore, cannot be left out of account.

When in March 1797 Bonaparte and the Army of Italy advanced

from the Tagliamento to meet the Archduke Charles,* their object was to force a decision on the Austrians* before the arrival of their reinforcements from the Rhine. If we consider only the immediate objective, the means were well-chosen, as the result showed. The Archduke's forces were still so weak that he made only an attempt at resistance on the Tagliamento. On seeing the strength and resolution of his enemy, he abandoned the area and the approaches to the Norican Alps. How could Bonaparte make use of this success? Should he press on into the heart of the Austrian Empire, ease the advance of the two armies of the Rhine under Moreau and Hoche,* and work in close conjunction with them? That was how Bonaparte saw it, and from his point of view he was right. But the critic may take a wider view—that of the French Directory;* whose members could see, and must have realized, that the campaign on the Rhine would not begin for another six weeks. From that standpoint, then, Bonaparte's advance through the Norican Alps could only be considered an unjustifiable risk. If the Austrians had moved sizeable reserves from the Rhine to Styria with which the Archduke Charles could have attacked the Army of Italy, not only would that Army have been destroyed, but the entire campaign would have been lost. Bonaparte realized this by the time he reached Villach, and this persuaded him to sign the Armistice of Leoben* with alacrity.

If the critic takes a still wider view, he can see that the Austrians had no reserves between the Archduke's army and Vienna, and that the advance of the Army of Italy was a threat to the capital itself.

Let us assume that Bonaparte knew the capital to be vulnerable and his own superiority over the Archduke even in Styria to be decisive. His rapid advance into the heart of Austria would then no longer have been pointless. The value of the attack would now depend merely on the value the Austrians set on the retention of Vienna. If, rather than lose the capital, they would accept whatever conditions for peace Bonaparte offered them, the threat to Vienna could be considered as his final aim. If Bonaparte had somehow known of this, the critic would have no more to say. But if the issue was still uncertain, the critic must take a more comprehensive point of view, and ask what would have happened if the Austrians had abandoned Vienna, and withdrawn into the vast expanse of territory they still controlled. That, however, is obviously a question which cannot possibly be answered without reference to the probable encounter between the two armies on the Rhine. There the French were so decisively superior in numbers—130,000 against 80,000—that the issue would not have been much in doubt. But then the question would again have arisen, what use would the French Directory have made of the victory? Would the French have pursued their advantage to the far frontiers of the Austrian monarchy, breaking Austrian power and shattering the Empire, or would they have been satisfied with the conquest of a sizeable part of it as a surety for peace? We have to ascertain the probable consequences of both possibilities before determining the probable choice of the Directory. Let us assume that this consideration led to the answer that the French forces were far too weak to bring about the total collapse of Austria, so that the mere attempt to do so would have reversed the situation and even the conquest and occupation of a significant segment of Austrian territory would have placed the French in a strategic situation with which their forces could hardly have coped. This argument would have coloured their view of the situation in which the Army of Italy found itself, and reduced its likely prospects. No doubt this is what persuaded Bonaparte, although he realized the Archduke's hopeless situation, to sign the peace of Campo Formio,* on conditions that imposed on the Austrians no greater sacrifices than the loss of some provinces which even the most successful campaign could not have recovered. But the French could not have counted even on the moderate gains of Campo Formio, and therefore could not have made them the objectives of their offensive, had it not been for two considerations. The first was the value the Austrians placed on the two possible outcomes. Though both of them made eventual success appear probable, would the Austrians have thought it worth the sacrifices they entailed—the continuation of the war—when that price could have been avoided by concluding a peace on not too unfavourable terms? The second consideration consists in the question whether the Austrian government would even pursue its reflections and thoroughly evaluate the potential limits of French success, rather than be disheartened by the impression of current reverses?

The first of these considerations is not simply idle speculation. On the contrary, it is of such decisive practical importance that it always arises whenever one aims at total victory. It is this which usually prevents such plans from being carried out.

The second consideration is just as essential, for war is not waged against an abstract enemy, but against a real one who must always be kept in mind. Certainly a man as bold as Bonaparte was conscious of this, confident as he was in the terror inspired by his approach. The same confidence led him to Moscow in 1812, but there it left him. In the course of the gigantic battles, the terror had already been somewhat blunted. But in 1797 it was still fresh, and the secret of the effectiveness of resisting to the last had not yet been discovered. Still, even in 1797 his boldness would have had a negative result if he had not, as we have seen, sensed the risk involved and chosen the moderate peace of Campo Formio as an alternative.

We must now break off this discussion. It will suffice to show the comprehensive, intricate and difficult character which a critical analysis may assume if it extends to ultimate objectives—in other words, if it deals with the great and decisive measures which must necessarily lead up to them. It follows that in addition to theoretical insight into the subject, natural talent will greatly enhance the value of critical analysis: for it will primarily depend on such talent to illuminate the connections which link things together and to determine which among the countless concatenations of events are the essential ones.

But talent will be needed in another way as well. Critical analysis is not just an evaluation of the means actually employed, but of *all possible means*—which first have to be formulated, that is, invented. One can, after all, not condemn a method without being able to suggest a better alternative. No matter how small the range of possible combinations may be in most cases, it cannot be denied that listing those that have not been used is not a mere analysis of existing things but an achievement that cannot be performed to order since it depends on the creativity of the intellect.

We are far from suggesting that the realm of true genius is to be found in cases where a handful of simple, practical schemes account for everything. In our view it is quite absurd, though it is often done, to treat the turning of a position as an invention of great genius. And yet such individual creative evaluations are necessary, and they significantly influence the value of critical analysis.

When on 30 July 1796, Bonaparte decided to raise the siege of Mantua* in order to meet Wurmser's* advance, and fell with his entire strength on each of the latter's columns separately while they were divided by Lake Garda and the Mincio, he did so because this seemed the surest way to decisive victories. These victories in fact did occur, and were repeated even more decisively in the same way against later attempts to relieve Mantua. There is only one opinion about this: unbounded admiration.

And yet, Bonaparte could not choose this course on 30 July without renouncing all hope of taking the city; for it was impossible to save the siege train, and it could not be replaced during the current campaign. In point of fact, the siege turned into a mere blockade and the city, which would have fallen within a week if the siege had been maintained, held out for six more months despite all Bonaparte's victories in the field.

Critics, unable to recommend a better way of resistance, have considered this an unavoidable misfortune. Resisting a relieving army behind lines of circumvallation had fallen into such disrepute and contempt that it occurred to no one. And yet in the days of Louis XIV* it had so often been successfully employed that one can only call it a whim of fashion that a hundred years later it never occurred to anyone *at least to weigh* its merits. If that possibility had been admitted, closer scrutiny of the situation would have shown that 40,000 of the finest infantrymen in the world whom Bonaparte could have placed behind a line of circumvallation at Mantua, would, if they were well entrenched, have had so little cause to fear the 50,000 Austrians whom Wurmser was bringing to relieve the town, that the lines were in little danger even of being attacked. This is not the place to labour the point; we believe we have said enough to show that the possibility deserved notice. We cannot tell whether Bonaparte himself ever considered the plan. There is no trace of it in his memoirs and the rest of the published sources; none of the later critics touched upon it, because they were no longer in the habit of considering this scheme. There is no great merit in recalling its existence; one only has to shed the tyranny of fashion in order to think of it. One does, however, have to think of it in order to consider it and to compare it with the means which Bonaparte in fact employed. Whatever the result of this comparison the critic should not fail to make it.

The world was filled with admiration when Bonaparte, in February 1814, turned from Blücher after beating him at Étoges, Champ-Aubert, Montmirail, and elsewhere, to fall on Schwarzenberg, and beat him at Montereau and Mormant.* By rapidly moving his main force back and forth, Bonaparte brilliantly exploited the allies' mistake of advancing with divided forces. If, people thought, these superb strokes in all directions failed to save him, at least it was not his fault. No one has yet asked what would have happened if, instead of turning away from Blücher, and back to Schwarzenberg, he had gone on hammering Blücher and had pursued him back to the Rhine. We are convinced that the complexion of the whole campaign would have been changed and that, instead of marching on Paris, the allied armies would have withdrawn across the Rhine. We do not require others to share our view, but no expert can doubt that the critic is bound to consider that alternative once it has been raised.

The option is much more obvious in this case than in the previous one. Nevertheless it has been overlooked, because people are biased and blindly follow a single line of thought.

The need for suggesting a better method than the one that is condemned has created the type of criticism which is used almost exclusively: the critic thinks he must only indicate the method which he considers to be better, without having to furnish proof. In consequence not everyone is convinced; others follow the same procedure, and a controversy starts without any basis for discussion. The whole literature on war is full of this kind of thing.

The proof that we demand is needed whenever the advantage of the means suggested is not plain enough to rule out all doubts; it consists in taking each of the means and assessing and comparing the particular merits of each in relation to the objective. Once the matter has thus been reduced to simple truths, the controversy must either stop, or at least lead to new results. By the other method, the pros and cons simply cancel out.

Suppose, for instance, that in the case of the last example, we had not been satisfied, and wanted to prove that the relentless pursuit of Blücher would have served Napoleon better than turning against Schwarzenberg. We would rely on the following simple truths:

1. Generally speaking, it is better to go on striking in the same direction than to move one's forces this way and that, because shifting troops back and forth involves losing time. Moreover, it is easier to achieve further successes where the enemy's morale has already been shaken by substantial losses; in this

way, none of the superiority that has been attained will go unexploited.

2. Even though Blücher was weaker than Schwarzenberg, his enterprising spirit made him more important. The centre of gravity lay with him, and he pulled the other forces in his direction.

3. The losses Blücher suffered were on the scale of a serious defeat. Bonaparte had thus gained so great a superiority over him as to leave no doubt that he would have to retreat as far as the Rhine, for no reserves of any consequence were stationed on that route.

4. No other possible success could have caused so much alarm or so impressed the allies' mind. With a staff which was known to be as timid and irresolute as Schwarzenberg's, this was bound to be an important consideration. The losses incurred by the Crown Prince of Württemberg* at Montereau and by Count Wittgenstein* at Mormant were sure to be fairly well known to Prince Schwarzenberg; on the other hand, news of the misfortunes that Blücher met with along his distant and discontinuous line between the Marne and the Rhine could have reached him only as an avalanche of rumours. Bonaparte's desperate thrust toward Vitry at the end of March was an attempt to test the effect that the threat of a strategic envelopment would have on the allies. It was obviously based on the principle of terror, but in wholly different circumstances now that Bonaparte had been defeated at Laon and Arcis,* and Blücher had joined Schwarzenberg with 100,000 men.

Some people, of course, will not be convinced by these arguments, but at least they will not be able to reply that 'as Bonaparte, in his thrust towards the Rhine, was threatening Schwarzenberg's base, so Schwarzenberg was threatening Paris, which was Bonaparte's'. The reasons we have cited above should make it clear that it would not have occurred to Schwarzenberg to advance on Paris.

In the instance from 1796 which we have touched on above we would say that Bonaparte considered the plan that he adopted as the one best guaranteed to beat the Austrians. Even if this had been true, the outcome would have been an empty triumph which could hardly have significantly affected the fall of Mantua. Our own proposal would have been much more likely to prevent Mantua from being relieved; but even if we put ourselves in Bonaparte's place and take the opposite view—that it offered a smaller prospect of success—the choice would have been based on balancing a likelier but almost useless, and therefore minor, victory against a less likely but far greater one. If the matter is looked at in that light, boldness would surely have opted for the second course: but looked at superficially, the opposite was what occurred. Bonaparte certainly held to the bolder intention, so there can be no doubt that he did not think the matter through to the point where he could assess the consequences as fully as we can in the light of experience.

In the study of means, the critic must naturally frequently refer to military history, for in the art of war experience counts more than any amount of abstract truths. Historical proof is subject to conditions of its own, which will be dealt with in a separate chapter; but unfortunately these conditions are so seldom met with that historical references usually only confuse matters more.

Another important point must now be considered: how far is the critic free, or even duty-bound, to assess a single case in the light of his greater knowledge, including as it does a knowledge of the outcome? Or when and where should he ignore these things in order to place himself exactly in the situation of the man in command?

If the critic wishes to distribute praise or blame, he must certainly try to put himself exactly in the position of the commander; in other words, he must assemble everything the commander knew and all the motives that affected his decision, and ignore all that he could not or did not know, especially the outcome. However, this is only an ideal to be aimed at, if never fully achieved: a situation giving rise to an event can never look the same to the analyst as it did to the participant. A mass of minor circumstances that may have influenced his decision are now lost to us, and many subjective motives may never have been exposed at all. These can only be discovered from the memoirs of the commanders, or from people very close to them. Memoirs often treat such matters pretty broadly, or, perhaps deliberately, with something less than candour. In short, the critic will always lack much that was present in the mind of the commander.

But it is even more difficult for the critic to shut off his superfluous knowledge. That is possible only with regard to accidental factors that impinge on the situation without being basic to it; in all really essential matters, however, it is very difficult and never fully attainable.

Let us first consider the outcome. Unless this was the result of chance, it is almost impossible to prevent the knowledge of it from colouring one's judgement of the circumstances from which it arose: we see these things in the light of their result, and to some extent come to know and appreciate them fully only because of it. Military history in all its aspects is itself a *source of instruction* for the critic, and it is only natural that he should look at all particular events in the light of the whole. Therefore, even if in some cases he did try to disregard results altogether, he could never entirely succeed.

But this is true not only of the outcome (that is, with what happens subsequently) but also of facts that were present from the beginning—the factors that determine the action. The critic will, as a rule, have more information than the participant. One would think he could easily ignore it, but he cannot. This is because knowledge of previous and simultaneous circumstances does not rest on specific information alone but on numerous conjectures and assumptions. Completely accidental matters apart, very little information does come to hand which has not been preceded by assumptions or conjectures. If specifics do not materialize, these assumptions and conjectures will take their place. Now we can understand why later critics who know all the previous and attendant circumstances must not be influenced by their knowledge when they ask which among the unknown facts they themselves would have considered probable at the time of the action. We maintain that complete insulation is as impossible here as it is when we consider the final outcome, and for the same reasons.

Therefore, if a critic wishes to praise or blame any specific action, he will only partly be able to put himself in the situation of the participant. In many cases he can do this well enough to suit practical purposes, but we must not forget that sometimes it is completely impossible.

It is, however, neither necessary nor desirable for the critic to identify himself completely with the commander. In war, as in all skills, a trained natural aptitude is called for. This virtuosity may be great or small. If it is great, it may easily be superior to that of the critic: what student would lay claim to the talent of a Frederick or a Bonaparte? Hence, unless we are to hold our peace in deference to outstanding talent, we must be allowed to profit from the wider horizons available to us. A critic should therefore not check a great commander's solution to a problem as if it were a sum in arithmetic. Rather, he must recognize with admiration the commander's success, the smooth unfolding of events, the higher workings of his genius. The essential interconnections that genius had divined, the critic has to reduce to factual knowledge.

To judge even the slightest act of talent, it is necessary for the critic to take a more comprehensive point of view, so that he, in possession of any number of objective reasons, reduces subjectivity to the minimum, and so avoids judging by his own, possibly limited, standards.

This elevated position of criticism, dispensing praise or blame with a full knowledge of all the circumstances, will not insult our feelings. The critic will do this only if he pushes himself into the limelight and implies that all the wisdom that is in fact derived from his complete knowledge of the case is due to his own abilities. No matter how crass that fraud, vanity may very easily lead to it, and it will naturally give offence. More often the critic does not mean to be arrogant; but, unless he makes a point of denying it, a hasty reader will suspect him of it, and this will at once give rise to a charge of lack of critical judgement.

If the critic points out that a Frederick or a Bonaparte made mistakes, it does not mean that he would not have made them too. He may even admit that in the situation of these generals he might have made far greater errors. What it does mean is that he can recognize these mistakes from the pattern of events and feels that the commander's sagacity should have seen them as well.

This is a judgement based on the pattern of events and therefore also *on their outcome*. But, in addition, the outcome may have a completely different effect on judgement—when the outcome is simply used as proof that an action was either correct or incorrect. This may be called a judgement *by results*. At first sight such a judgement would seem entirely inadmissible, but that is not the case.

When in 1812 Bonaparte advanced on Moscow* the crucial question was whether the capture of the capital, together with everything else that had already happened, would induce Czar Alexander* to make peace. That had happened in 1807 after the battle of Friedland,* and it had also worked in 1805 and 1809 with the Emperor Francis* after the battles of Austerlitz and Wagram.* If, however, peace was not made at Moscow, Bonaparte would have no choice but to turn back, which would have meant a strategic defeat. Let us leave aside the steps by which he advanced on Moscow, and the question whether, in the process, he missed a number of opportunities that might have made the Czar decide on peace. Let us also leave aside the terrible circumstances of the retreat, which may have had their root in the conduct of the entire campaign. The crucial question remains the same: no matter how much more successful the advance on Moscow might have been, it would still have been uncertain whether it could have frightened the Czar into suing for peace. And even if the retreat had not led to the annihilation of the army, it could never have been anything but a major strategic defeat. If the Czar had concluded a disadvantageous peace, the campaign of 1812 would have ranked with those of Austerlitz, Friedland, and Wagram. But if these campaigns had not resulted in peace, they would probably have led to similar catastrophes. Regardless of the power, skill, and wisdom shown by the conqueror of the world, the final fatal question remained everywhere the same. Should we then ignore the actual results of the campaigns of 1805, 1807, and 1809, and, by the test of 1812 alone, proclaim them to be products of imprudence, and their success to be a breach of natural law? Should we maintain that in 1812 strategic justice finally overcame blind chance? That would be a very forced conclusion, an arbitrary judgement where half the evidence is missing, because the human eye cannot trace the interconnection of events back to the decisions of the vanquished monarchs.

Still less can it be said that the campaign of 1812 ought to have succeeded like the others, and that its failure was due to something extraneous: there was nothing extraneous about Alexander's steadfastness.

What can be more natural than to say that in 1805, 1807, and 1809 Bonaparte had gauged his enemy correctly, while in 1812 he did not? In the earlier instances he was right, in the latter he was wrong, and we can say that *because the outcome proves it*.

In war, as we have already pointed out, all action is aimed at probable rather than at certain success. The degree of certainty that is lacking must in every case be left to fate, chance, or whatever you like to call it. One may of course ask that this dependence should be as slight as possible, but only in reference to a particular case—in other words, it should be as *small as possible in that individual case*. But we should not habitually prefer the course that involves the least uncertainty. That would be an enormous mistake, as our theoretical arguments will show. There are times when the utmost daring is the height of wisdom.

It would seem that a commander's personal merits, and thus also his responsibility, become irrelevant to all questions that have to be left to chance. Nevertheless, we cannot deny an inner satisfaction whenever things turn out right; when they do not, we feel a certain intellectual discomfort. *That is all the meaning that should be attached to a judgement of right and wrong that we deduce from success, or rather that we find in success.*

But it is obvious that the intellectual pleasure at success and the intellectual discomfort at failure arise from an obscure sense of some delicate link, invisible to the mind's eye, between success and the commander's genius. It is a gratifying assumption. The truth of this is shown by the fact that our sympathy increases and grows keener as success and failure are repeated by the same man. That is why luck in war is of higher quality than luck in gambling. So long as a successful general has not done us any harm, we follow his career with pleasure.

The critic, then, having analysed everything within the range of human calculation and belief, will let the outcome speak for that part whose deep, mysterious operation is never visible. The critic must protect this unspoken result of the workings of higher laws against the stream of uninformed opinion on the one hand, and against the gross abuses to which it may be subjected on the other.

Success enables us to understand much that the workings of human intelligence alone would not be able to discover. That means that it will be useful mainly in revealing intellectual and psychological forces and effects, because these are least subject to reliable evaluation, and also because they are so closely involved with the will that they may easily control it. Wherever decisions are based on fear or courage, they can no longer be judged objectively; consequently, intelligence and calculation can no longer be expected to determine the probable outcome.

We must now be allowed to make a few remarks about the instruments critics use—their idiom; for in a sense it accompanies action in war. Critical analysis, after all, is nothing but thinking that should precede the action. We therefore consider it essential that the language of criticism should have the same character as thinking must have in wars; otherwise it loses its practical value and criticism would lose contact with its subject.

In our reflections on the theory of the conduct of war, we said that it ought to train a commander's mind, or rather, guide his education; theory is not meant to provide him with positive doctrines and systems to be used as intellectual tools. Moreover, if it is never necessary or even permissible to use scientific guidelines in order to judge a given problem in war, if the truth never appears in systematic form, if it is not acquired deductively but always *directly* through the natural perception of the mind, then that is the way it must also be in critical analysis.

We must admit that wherever it would be too laborious to determine the facts of the situation, we must have recourse to the relevant principles established by theory. But in the same way as in war these truths are better served by a commander who has absorbed their meaning in his mind rather than one who treats them as rigid external rules, so the critic should not apply them like an external law or an algebraic formula whose relevance need not be established each time it is used. These truths should always be allowed to become self-evident, while only the more precise and complex proofs are left to theory. We will thus avoid using an arcane and obscure language, and express ourselves in plain speech, with a sequence of clear, lucid concepts.

Granted that while this cannot always be completely achieved, it must remain the aim of critical analysis. The complex forms of cognition should be used as little as possible, and one should never use elaborate scientific guidelines as if they were a kind of truth machine. Everything should be done through the natural workings of the mind.

However, this pious aspiration, if we may call it that, has rarely prevailed in critical studies; on the contrary, a kind of vanity has impelled most of them to an ostentatious exhibition of ideas.

The first common error is an awkward and quite impermissible use of certain narrow systems as formal bodies of laws. It is never difficult to demonstrate the one-sidedness of such systems; and nothing more is needed to discredit their authority once and for all. We are dealing here with a limited problem, and since the number of possible systems is after all finite, this error is the lesser of two evils that concern us.

A far more serious menace is the retinue of *jargon, technicalities, and metaphors* that attends these systems. They swarm everywhere—a lawless rabble of camp followers. Any critic who has not seen fit to adopt a system—either because he has not found one that he likes or because he has not yet got that far—will still apply an occasional scrap of one as if it were a ruler, to show the crookedness of a commander's course. Few of them can proceed without the occasional support of such scraps of scientific military theory. The most insignificant of them—mere technical expressions and metaphors—are sometimes nothing more than ornamental flourishes of the critical narrative. But it is inevitable that all the terminology and technical expressions of a given system will lose what meaning they have, if any, once they are torn from their context and used as general axioms or nuggets of truth that are supposed to be more potent than a simple statement.

Thus it has come about that our theoretical and critical literature, instead of giving plain, straightforward arguments in which the author at least always knows what he is saying and the reader what he is reading, is crammed with jargon, ending at obscure crossroads where the author loses his readers. Sometimes these books are even worse: they are just hollow shells. The author himself no longer knows just what he is thinking and soothes himself with obscure ideas which would not satisfy him if expressed in plain speech.

Critics have yet a third failing: showing off their erudition, and the misuse of historical examples. We have already stated what the history of the art of war is, and our views on historical examples and military history in general will be developed in later chapters. A fact that is cited in passing may be used to support *the most contradictory views*; and three or four examples from distant times and places, dragged in and piled up from the widest range of circumstances, tend to distract and confuse one's judgement without proving anything. The light of day usually reveals them to be mere trash, with which the author intends to show off his learning.

What is the practical value of these obscure, partially false, confused and arbitrary notions? Very little—so little that they have made theory, from its beginnings, the very opposite of practice, and not infrequently the laughing stock of men whose military competence is beyond dispute.

This could never have happened if by means of simple terms and straightforward observation of the conduct of war theory had sought to determine all that was determinable; if, without spurious claims, with no unseemly display of scientific formulae and historical compendia, it had stuck to the point and never parted company with those who have to manage things in battle by the light of their native wit.

CHAPTER 6
ON HISTORICAL EXAMPLES

HISTORICAL examples clarify everything and also provide the best kind of proof in the empirical sciences. This is particularly true of the art of war. General Scharnhorst,* whose manual is the best that has ever been written about actual war, considers historical examples to be of prime importance to the subject, and he makes admirable use of them. If he had survived the wars of 1813–1815, the fourth part of his revised work on artillery would have demonstrated even better the powers of observation and instruction with which he treated his experiences.

Historical examples are, however, seldom used to such good effect. On the contrary, the use made of them by theorists normally not only leaves the reader dissatisfied but even irritates his intelligence. We therefore consider it important to focus attention on the proper and improper uses of examples.

Undoubtedly, the knowledge basic to the art of war is empirical. While, for the most part, it is derived from the nature of things, this very nature is usually revealed to us only by experience. Its application, moreover, is modified by so many conditions that its effects can never be completely established merely from the nature of the means.

The effects of gunpowder—that major agent of military activity—could only be demonstrated by experience. Experiments are still being conducted to study them more closely.

It is, of course, obvious that an iron cannonball, impelled by powder to a speed of 1,000 feet per second, will smash any living creature in its path. One needs no experience to believe that. But there are hundreds of relevant details determining this effect, some of which can only be revealed empirically. Nor is the physical effect the only thing that matters: the psychological effect is what concerns us, and experience is the only means by which it can be established and appreciated. In the Middle Ages firearms were a new invention, so crude that their physical effect was much less important than today; but their psychological impact was considerably greater. One has to have seen the steadfastness of one of the forces trained and led by Bonaparte in the course of his conquests—seen them under fierce and unrelenting fire—to get some sense of what can be accomplished by troops steeled by long experience of danger, in whom a proud record of victories has instilled the noble principle of placing the highest demands on themselves. As an idea alone it is unbelievable. On the other hand, there are European armies that still have troops such as Tartars, Cossacks, and Croats* whose ranks can easily be scattered by a few rounds of artillery.

Still, the empirical sciences, the theory of the art of war included, cannot always back their conclusions with historical proofs. The sheer range to be covered would often rule this out; and, apart from that, it might be difficult to point to actual experience on every detail. If, in warfare, a certain means turns out to be highly effective, it will be used again; it will be copied by others and become fashionable; and so, backed by experience, it passes into general use and is included in theory. Theory is content to refer to experience in general to indicate the origin of the method, but not to prove it.

It is a different matter when experience is cited in order to displace a method in current usage, confirm a dubious, or introduce a new one. In those cases, individual instances from history must be produced as evidence.

A closer look at the use of historical examples will enable us to distinguish four points of view.

First, a historical example may simply be used as an *explanation* of an idea. Abstract discussion, after all, is very easily misunderstood, or not understood at all. When an author fears that this might happen, he may use a historical example to throw the necessary light on his idea and to ensure that the reader and the writer will remain in touch.

Second, it may serve to show the *application* of an idea. An example gives one the opportunity of demonstrating the operation of all those minor circumstances which could not be included in a general formulation of the idea. Indeed, this is the difference between theory and experience. Both the foregoing cases concerned true examples; the two that follow concern historical proof.

Third, one can appeal to historical fact to support a statement. This will suffice wherever one merely wants to prove the *possibility* of some phenomenon or effect.

Fourth and last, the detailed presentation of a historical event, and the combination of several events, make it possible to deduce a doctrine: the proof is in the evidence itself.

The use of the first type generally calls only for a brief mention of the case, for only one aspect of it matters. Historical truth is not even essential here: an imaginary case would do as well. Still, historical examples always have the advantage of being more realistic and of bringing the idea they are illustrating to life.

The second type of usage demands a more detailed presentation of events; but authenticity, once again, is not essential. In this respect, we repeat what we said about the first case.

The third purpose is sufficiently met, as a rule, by the simple statement of an undisputed fact. If one is trying to show that an entrenched position can under certain circumstances prove effective, a mention of the Bunzelwitz* position will support the statement.

If, however, some historical event is being presented in order to demonstrate a general truth, care must be taken that every aspect bearing on the truth at issue is fully and circumstantially developed—carefully assembled, so to speak, before the reader's eyes. To the extent that this cannot be done, the proof is weakened, and the more necessary it will be to use a number of cases to supply the evidence missing in that one. It is fair to assume that where we cannot cite more precise details, the average effect will be decided by a greater number of examples.

Suppose one wants to prove from experience that cavalry should be placed in the rear of infantry rather than in line with it; or that, without definite numerical superiority, it is extremely dangerous to use widely separated columns in attempting to envelop the enemy, both on the battlefield and in the theatre of operations—tactically or strategically, in other words. As to the first instance, it is not enough to cite a few defeats where the cavalry was on the flanks, and a few victories where it was behind the infantry; in the second case, it will not be enough to refer to the battles of Rivoli or Wagram, and the Austrian attacks on the Italian theatre, or those of the French on the German theatre of war, in 1796. Instead one must accurately trace all the circumstances and individual events, to show the way in which those types of position and attack definitely contributed to the defeat. The result will show *to what degree* these types are objectionable—a point that must be settled in any case, because a general condemnation would conflict with the truth.

We have already agreed that where a detailed factual account cannot be given, any lack of evidence may be made up by the number of examples; but this is clearly a dangerous expedient, and is frequently misused. Instead of presenting a fully detailed case, critics are content merely to *touch on* three or four, which give the *semblance* of strong proof. But there are occasions where nothing will be proved by a dozen examples—if, for instance, they frequently recur and one could just as easily cite a dozen cases that had opposite results. If anyone lists a dozen defeats in which the losing side attacked with divided columns, I can list a dozen victories in which that very tactic was employed. Obviously this is no way to reach a conclusion.

Reflection upon these diverse circumstances will show how easily examples may be misused.

An event that is lightly touched upon, instead of being carefully detailed, is like an object seen at a great distance: it is impossible to distinguish any detail, and it looks the same from every angle. Such examples have actually been used to support the most conflicting views. Daun's campaigns* are, to some, models of wisdom and foresight; to others, of timidity and vacillation. Bonaparte's thrust across the Norican Alps in 1797 strikes some as a splendid piece of daring; others will call it completely reckless. His strategic defeat in 1812 may be put down to an excess of energy, but also to a lack of it. All these views have been expressed, and one can easily see why: the pattern of events was interpreted in different ways. Nevertheless, these conflicting opinions cannot coexist; one or the other must be wrong.

Feuquières,* that excellent man, deserves our thanks for the wealth of examples that adorn his memoirs. He not only records a number of events that would otherwise have been forgotten; he was the first to make really useful comparisons between abstract theoretical ideas and real life insofar as the cases cited can be considered as explanations and closer definitions of his theoretical assertions. Still, to an impartial modern reader, he has hardly achieved the aim he usually set himself, that of proving theoretical principles by historical examples. Though he occasionally records events in some detail, he still falls short of proving that the conclusions he has drawn are the inevitable consequences of their inherent patterns.

Another disadvantage of merely touching on historical events lies in the fact that some readers do not know enough about them, or do not remember them well enough to grasp what the author has in mind. Such readers have no choice but to be impressed by the argument, or to remain untouched by it altogether.

It is hard of course to recount a historical event or reconstruct it for the reader in the way required if it is to be used as evidence. The writer rarely has the means, the space, or the time for that. We maintain, however, that where a new or debatable point of view is concerned, a single thoroughly detailed event is more instructive than ten that are only touched on. The main objection to this superficial treatment is not that the writer pretends he is trying to prove something but that he himself has never mastered the events he cites, and that such superficial, irresponsible handling of history leads to hundreds of wrong ideas and bogus theorizing. None of this would come about if the writer's duty were to show that the new ideas he is presenting as guaranteed by history are indisputably derived from the precise pattern of events.

Once one accepts the difficulties of using historical examples, one will come to the most obvious conclusion that examples should be drawn from modern military history, insofar as it is properly known and evaluated.

Not only were conditions different in more distant times, with different ways of waging war, so that earlier wars have fewer practical lessons for us; but military history, like any other kind, is bound with the passage of time to lose a mass of minor elements and details that were once clear. It loses some element of life and colour, like a picture that gradually fades and darkens. What remains in the end, more or less at random, are large masses and isolated features, which are thereby given undue weight.

If we examine the conditions of modern warfare, we shall find that the wars that bear a considerable resemblance to those of the present day, especially with respect to armaments, are primarily campaigns beginning with the War of the Austrian Succession.* Even though many major and minor circumstances have changed considerably, these are close enough to modern warfare to be instructive. The situation is different with the War of the Spanish Succession;* the use of firearms was much less advanced, and cavalry was still the most important arm. The further back one goes, the less useful military history becomes, growing poorer and barer at the same time. The history of antiquity is without doubt the most useless and the barest of all.

This uselessness is of course not absolute; it refers only to matters that depend on a precise knowledge of the actual circumstances, or on details in which warfare has changed. Little as we may know about the battles the Swiss fought against the Austrians, the Burgundians, and the French, it is they that afford the first and strongest demonstration of the superiority of good infantry against the best cavalry. A general glance at the age of the *condottieri** is enough to show that the conduct of war depends entirely on the instrument employed; at no other time were the forces used so specialized in character or so completely divorced from the rest of political and civil life. The peculiar way in which Rome fought Carthage in the Second Punic War—by attacking Spain and Africa while Hannibal* was still victorious in Italy—can provide a most instructive lesson: we still know enough about the general situation of the states and armies that enabled such a roundabout method of resistance to succeed.

But the further one progresses from broad generalities to details, the less one is able to select examples and experiences from remote times. We are in no position to evaluate the relevant events correctly, nor to apply them to the wholly different means we use today.

Unfortunately, writers have always had a pronounced tendency to refer to events in ancient history. How much of this is due to vanity and quackery can remain unanswered; but one rarely finds any honesty of purpose, any earnest attempt to instruct or convince. Such allusions must therefore be looked upon as sheer decoration, designed to cover gaps and blemishes.

To teach the art of war entirely by historical examples, which is what Feuquières tried to do, would be an achievement of the utmost value; but it would be more than the work of a lifetime: anyone who set out to do it would first have to equip himself with a thorough personal experience of war.

Anyone who feels the urge to undertake such a task must dedicate himself for his labours as he would prepare for a pilgrimage to distant lands. He must spare no time or effort, fear no earthly power or rank, and rise above his own vanity or false modesty in order to tell, in accordance with the expression of the *Code Napoléon,** the truth, the whole truth, and nothing but the truth.*

BOOK THREE
ON STRATEGY IN GENERAL

STRATEGY

THE general concept of strategy was defined in the second chapter of Book Two.* It is the use of an engagement for the purpose of the war. Though strategy in itself is concerned only with engagements, the theory of strategy must also consider its chief means of execution, the fighting forces. It must consider these in their own right and in their relation to other factors, for they shape the engagement and it is in turn on them that the effect of the engagement first makes itself felt. Strategic theory must therefore study the engagement in terms of its possible results and of the moral and psychological forces that largely determine its course.

Strategy is the use of the engagement for the purpose of the war. The strategist must therefore define an aim for the entire operational side of the war that will be in accordance with its purpose. In other words, he will draft the plan of the war, and the aim will determine the series of actions intended to achieve it: he will, in fact, shape the individual campaigns and, within these, decide on the individual engagements. Since most of these matters have to be based on assumptions that may not prove correct, while other, more detailed orders cannot be determined in advance at all, it follows that the strategist must go on campaign himself. Detailed orders can then be given on the spot, allowing the general plan to be adjusted to the modifications that are continuously required. The strategist, in short, must maintain control throughout.

This has not always been the accepted view, at least so far as the general principle is concerned. It used to be the custom to settle strategy in the capital, and not in the field—a practice that is acceptable only if the government stays so close to the army as to function as general headquarters.

Strategic theory, therefore, deals with planning; or rather, it attempts to shed light on the components of war and their interrelationships, stressing those few principles or rules that can be demonstrated.

The reader who recalls from the first chapter of Book I how many vitally important matters are involved in war will understand what unusual mental gifts are needed to keep the whole picture steadily in mind.

A prince or a general can best demonstrate his genius by managing a campaign exactly to suit his objectives and his resources, doing neither too much nor too little. But the effects of genius show not so much in novel forms of action as in the ultimate success of the whole. What we should admire is the accurate fulfilment of the unspoken assumptions, the smooth harmony of the whole activity, which only become evident in final success.

The student who cannot discover this harmony in actions that lead up to a final success may be tempted to look for genius in places where it does not and cannot exist.

In fact, the means and forms that the strategist employs are so very simple, so familiar from constant repetition, that it seems ridiculous in the light of common sense when critics discuss them, as they do so often, with ponderous solemnity. Thus, such a commonplace manoeuvre as turning an opponent's flank may be hailed by critics as a stroke of genius, of deepest insight, or even of all-inclusive knowledge. Can one imagine anything more absurd?

It is even more ridiculous when we consider that these very critics usually exclude all moral qualities from strategic theory, and only examine material factors. They reduce everything to a few mathematical formulas of equilibrium and superiority, of time and space, limited by a few angles and lines. If that were really all, it would hardly provide a scientific problem for a schoolboy.

But we should admit that scientific formulas and problems are not under discussion. The relations between material factors are all very simple; what is more difficult to grasp are the intellectual factors involved. Even so, it is only in the highest realms of strategy that intellectual complications and extreme diversity of factors and relationships occur. At that level there is little or no difference between strategy, policy and statesmanship, and there, as we have already said, their influence is greater in questions of quantity and scale than in forms of execution. Where execution is dominant, as it is in the individual events of a war whether great or small, then intellectual factors are reduced to a minimum.

Everything in strategy is very simple, but that does not mean that everything is very easy. Once it has been determined, from the political conditions, what a war is meant to achieve and what it can achieve, it is easy to chart the course. But great strength of character, as well as great lucidity and firmness of mind, is required in order to follow through steadily, to carry out the plan, and not to be thrown off course by thousands of diversions. Take any number of outstanding men, some noted for intellect, others for their acumen, still others for boldness or tenacity of will: not one may possess the combination of qualities needed to make him a greater than average commander.

It sounds odd, but everyone who is familiar with this aspect of warfare will agree that it takes more strength of will to make an important decision in strategy than in tactics. In the latter, one is carried away by the pressures of the moment, caught up in a maelstrom where resistance would be fatal, and, suppressing incipient scruples, one presses boldly on. In strategy, the pace is much slower. There is ample room for apprehensions, one's own and those of others; for objections and remonstrations and, in consequence, for premature regrets. In a tactical situation one is able to see at least half the problem with the naked eye, whereas in strategy everything has to be guessed at and presumed. Conviction is therefore weaker. Consequently most generals, when they ought to act, are paralysed by unnecessary doubts.

Now a glance at history. Let us consider the campaign that Frederick the Great fought in 1760, famous for its dazzling marches and manoeuvres, praised by critics as a work of art—indeed a masterpiece. Are we to be beside ourselves with admiration at the fact that the King wanted first to turn Daun's right flank, then his left, then his right again, and so forth? Are we to consider this profound wisdom? Certainly not, if we are to judge without affectation. What is really admirable is the King's wisdom: pursuing a major objective with limited resources, he did not try to undertake anything beyond his strength, but always *just enough* to get him what he wanted. This campaign was not the only one in which he demonstrated his judgement as a general. It is evident in all the three wars fought by the great King.

His object was to bring Silesia into the safe harbour of a fully guaranteed peace.

As head of a small state resembling other states in most respects, and distinguished from them only by the efficiency of some branches of its administration, Frederick could not be an Alexander.* Had he acted like Charles XII,* he too would have ended in disaster. His whole conduct of war, therefore, shows an element of restrained strength, which was always in balance, never lacking in vigour, rising to remarkable heights in moments of crisis, but immediately afterward reverting to a state of calm oscillation, always ready to adjust to the smallest shift in the political situation. Neither vanity, ambition, nor vindictiveness could move him from this course; and it was this course alone that brought him success.

How little these few words can do to appreciate that characteristic of the great general! One only has to examine carefully the causes and the miraculous outcome of this struggle to realize that it was only the King's acute intelligence that led him safely through all hazards.

This is the characteristic we admire in all his campaigns, but especially in the campaign of 1760. At no other time was he able to hold off such a superior enemy at so little cost.

The other aspect to be admired concerns the difficulties of execution. Manoeuvres designed to turn a flank are easily planned. It is equally easy to conceive a plan for keeping a small force concentrated so that it can meet a scattered enemy on equal terms at any point, and to multiply its strength by rapid movement. There is nothing admirable about the ideas themselves. Faced with such simple concepts, we have to admit that they are simple.

But let a general try to imitate Frederick! After many years eye-witnesses still wrote about the risk, indeed the imprudence, of the King's positions; and there can be no doubt that the danger appeared three times as threatening at the time as afterward.

It was the same with the marches undertaken under the eyes, frequently under the very guns, of the enemy. Frederick chose these positions and made these marches, confident in the knowledge that Daun's methods, his dispositions, his sense of responsibility and his character would make such manoeuvres risky but not reckless. But it required the King's boldness, resolution, and strength of will to see things in this way, and not to be confused and intimidated by the danger that was still being talked and written about thirty years later. Few generals in such a situation would have believed such simple means of strategy to be feasible.

Another difficulty of execution lay in the fact that throughout this campaign the King's army was constantly on the move. Twice, in early July and early August, it followed Daun while itself pursued by Lacy,* from the Elbe into Silesia over wretched country roads. The army had to be ready for battle at any time, and its marches had to be organized with a degree of ingenuity that required a proportionate amount of exertion. Though the army was accompanied, and delayed, by thousands of wagons, it was always short of supplies. For a week before the battle of Liegnitz* in Silesia, the troops marched day and night, alternatively deploying and withdrawing along the enemy's front. This cost enormous exertions and great hardship.

Could all this be done without subjecting the military machine to serious friction? Is a general, by sheer force of intellect, able to produce such mobility with the ease of a surveyor manipulating an astrolabe? Are the generals and the supreme commander not moved by the sight of the misery suffered by their pitiful, hungry, and thirsty comrades in arms? Are complaints and misgivings about such conditions not reported to the high command? Would an ordinary man dare to ask for such sacrifices, and would these not automatically lower the morale of the troops, corrupt their discipline, in short undermine their fighting spirit unless an overwhelming belief in the greatness and infallibility of their commander outweighed all other considerations? It is this which commands our respect; it is these miracles of execution that we have to admire. But to appreciate all this in full measure one has to have had a taste of it through actual experience. Those who know war only from books or the parade-ground cannot recognize the existence of these impediments to action, and so we must ask them to accept on faith what they lack in experience.

We have used the example of Frederick to bring our train of thought into focus. In conclusion, we would point out that in our exposition of strategy we shall describe those material and intellectual factors that seem to us to be the most significant. We shall proceed from the simple to the complex, and conclude with the unifying structure of the entire military activity—that is, with the plan of campaign.

An earlier manuscript of Book Two contains the following passages, marked by the author: 'To be used in the first chapter of Book Three.' The projected revision of this chapter was never made, and these passages are therefore inserted here in full.

In itself, the deployment of forces at a certain point merely makes an engagement possible; it does not necessarily take place. Should one treat this possibility as a reality, as an actual occurrence? Certainly. It becomes real because of its consequences, and *consequences of some kind will always follow*.

Possible Engagements Are To Be Regarded As Real Ones Because of Their Consequences

If troops are sent to cut off a retreating enemy and he thereupon surrenders without further fight, his decision is caused solely by the threat of a fight posed by those troops.

If part of our army occupies an undefended enemy province and thus denies the enemy substantial increments to his strength, the factor making it possible for our force to hold the province is the engagement that the enemy must expect to fight if he endeavours to retake it.

In both cases results have been produced by the mere possibility of an engagement; the possibility has acquired reality. But let us suppose that in each case the enemy had brought superior forces against our troops, causing them to abandon their goal without fighting. This would mean that we had fallen short of our objective; but still the engagement that we offered the enemy was not without effect—it did draw off his forces. Even if the whole enterprise leaves us worse off than before, we cannot say that no effects resulted from using troops in this way, by producing the *possibility of an engagement*; the effects were similar to those of a lost engagement.

This shows that the destruction of the enemy's forces and the overthrow of the enemy's power can be accomplished only as the result of an engagement, no matter whether it really took place or was merely offered but not accepted.

The Twofold Object of the Engagement

These results, moreover, are of two kinds: direct and indirect. They are indirect if other things intrude and become the object of the engagement—things which cannot in themselves be considered to involve the destruction of the enemy's forces, but which lead up to it. They may do so by a circuitous route, but are all the more powerful for that. The possession of provinces, cities, fortresses, roads, bridges, munitions dumps, etc., may be the *immediate* object of an engagement, but can never be the final one. Such acquisitions should always be regarded merely as means of gaining greater superiority, so that in the end we are able to offer an engagement to the enemy when he is in no position to accept it. These actions should be considered as intermediate links, as steps leading to the operative principle, never as the operative principle itself.

Examples

With the occupation of Bonaparte's capital in 1814, the objective of the war had been achieved. The political cleavages rooted in Paris came to the surface, and that enormous split caused the Emperor's power to collapse. Still, all this should be considered in the light of the military implications. The occupation caused a substantial diminution in Bonaparte's military strength and his capacity to resist, and a corresponding increase in the superiority of the allies. Further resistance became impossible, and it was this which led to peace with France. Suppose the allied strength had suddenly been similarly reduced by some external cause: their superiority would have vanished, and with it the whole effect and significance of their occupation of Paris.

We have pursued this argument to show that this is the natural and only sound view to take, and this is what makes it important. We are constantly brought back to the question: what, at any given stage of the war or campaign, will be the likely outcome of all the major and minor engagements that the two sides can offer one another? In the planning of a campaign or a war, this alone will decide the measures that have to be taken from the outset.

If This View Is Not Adopted, Other Matters Will Be Inaccurately Assessed

If we do not learn to regard a war, and the separate campaigns of which it is composed, as a chain of linked engagements each leading to the next, but instead succumb to the idea that the capture of certain geographical points or the seizure of undefended provinces are *of value in themselves*, we are liable to regard them as windfall profits. In so doing, and in ignoring the fact that they are links in a continuous chain of events, we also ignore the possibility that their possession may later lead to definite disadvantages. This mistake is illustrated again and again in military history. One could almost put the matter this way: just as a businessman cannot take the profit from a single transaction and put it into a separate account, so an isolated advantage gained in war cannot be assessed separately from the over-all result. A businessman must work on the basis of his total assets, and in war the advantages and disadvantages of a single action could only be determined by the final balance.

By looking on each engagement as part of a series, at least insofar as events are predictable, the commander is always on the high road to his goal. The forces gather momentum, and intentions and actions develop with a vigour that is commensurate with the occasion, and impervious to outside influences.

CHAPTER 2

ELEMENTS OF STRATEGY

THE strategic elements that affect the use of engagements may be classified into various types: moral, physical, mathematical, geographical, and statistical.

The first type covers everything that is created by intellectual and psychological qualities and influences; the second consists of the size of the armed forces, their composition, armament and so forth; the third includes the angle of lines of operation, the convergent and divergent movements wherever geometry enters into their calcula-tion; the fourth comprises the influence of terrain, such as com-manding positions, mountains, rivers, woods, and roads; and, finally, the fifth covers support and maintenance. A brief consideration of each of these various types will clarify our ideas and, in passing, assess the relative value of each. Indeed if they are studied separately some will automatically be stripped of any undue importance. For instance, it immediately becomes clear that the value of the base of operations, even if we take this in its simplest form as meaning a *base-line*, depends less on its geometric forms than on the nature of the roads and terrain through which they run.

It would however be disastrous to try to develop our understanding of strategy by analysing these factors in isolation, since they are usually interconnected in each military action in manifold and intricate ways. A dreary analytical labyrinth would result, a nightmare in which one tried in vain to bridge the gulf between this abstract basis and the facts of life. Heaven protect the theorist from such an undertaking! For our part, we shall continue to examine the picture as a whole, and take our analysis no further than is necessary in each case to elucidate the idea we wish to convey, which will always have its origins in the impressions made by the sum total of the phenomena of war, rather than in speculative study.

CHAPTER 3

MORAL FACTORS

WE must return once more to this subject, already touched upon in Chapter Three of Book Two* since the moral elements are among the most important in war. They constitute the spirit that permeates war as a whole, and at an early stage they establish a close affinity with the will that moves and leads the whole mass of force, practically merging with it, since the will is itself a moral quantity. Unfortunately they will not yield to academic wisdom. They cannot be classified or counted. They have to be seen or felt.

The spirit and other moral qualities of an army, a general or a government, the temper of the population of the theatre of war, the moral effects of victory or defeat—all these vary greatly. They can moreover influence our objective and situation in very different ways.

Consequently, though next to nothing can be said about these things in books, they can no more be omitted from the theory of the art of war than can any of the other components of war. To repeat, it is paltry philosophy if in the old-fashioned way one lays down rules and principles in total disregard of moral values. As soon as these appear one regards them as exceptions, which gives them a certain scientific status, and thus makes them into rules. Or again one may appeal to genius, which is above all rules; which amounts to admitting that rules are not only made for idiots, but are idiotic in themselves.

If the theory of war did no more than remind us of these elements, demonstrating the need to reckon with and give full value to moral qualities, it would expand its horizon, and simply by establishing this point of view would condemn in advance anyone who sought to base an analysis on material factors alone.

Another reason for not placing moral factors beyond the scope of theory is their relation to all other so-called rules. The effects of physical and psychological factors form an organic whole which, unlike a metal alloy, is inseparable by chemical processes. In formulating any rule concerning physical factors, the theorist must bear in mind the part that moral factors may play in it; otherwise he may be misled into making categorical statements that will be too timid and restricted, or else too sweeping and dogmatic. Even the most uninspired theories have involuntarily had to stray into the area of intangibles; for instance, one cannot explain the effects of a victory without taking psychological reactions into account. Hence most of the matters dealt with in this book are composed in equal parts of physical and of moral causes and effects. One might say that the physical seem little more than the wooden hilt, while the moral factors are the precious metal, the real weapon, the finely-honed blade.

History provides the strongest proof of the importance of moral factors and their often incredible effect: this is the noblest and most solid nourishment that the mind of a general may draw from a study of the past. Parenthetically, it should be noted that the seeds of wisdom that are to bear fruit in the intellect are sown less by critical studies and learned monographs than by insights, broad impressions, and flashes of intuition.

We might list the most important moral phenomena in war and, like a diligent professor, try to evaluate them one by one. This method, however, all too easily leads to platitudes, while the genuine spirit of inquiry soon evaporates, and unwittingly we find ourselves proclaiming what everybody already knows. For this reason we prefer, here even more than elsewhere, to treat the subject in an incomplete and impressionistic manner, content to have pointed out its general importance and to have indicated the spirit in which the arguments of this book are conceived.

CHAPTER 4

THE PRINCIPAL MORAL ELEMENTS

THEY are: *the skill of the commander, the experience and courage of the troops, and their patriotic spirit.* The relative value of each cannot be universally established; it is hard enough to discuss their potential, and even more difficult to weigh them against each other. The wisest course is not to underrate any of them—a temptation to which human judgement, being fickle, often succumbs. It is far preferable to muster historical evidence of the unmistakable effectiveness of all three.

Nevertheless it is true that at this time the armies of practically all European states have reached a common level of discipline and training. To use a philosophic expression: the conduct of war has developed in accordance with its natural laws. It has evolved methods that are common to most armies, and that no longer even allow the commander scope to employ special artifices (in the sense, for example, of Frederick the Great's oblique order of battle). It cannot be denied, therefore, that as things stand at present proportionately greater scope is given to the troops' patriotic spirit and combat experience. A long period of peace may change this again.

The troops' national feeling (enthusiasm, fanatical zeal, faith, and general temper) is most apparent in mountain warfare where every man, down to the individual soldier, is on his own. For this reason alone mountainous areas constitute the terrain best suited for action by an armed populace.

Efficiency, skill, and the tempered courage that welds the body of troops into a single mould will have their greatest scope in operations in open country.

The commander's talents are given greatest scope in rough hilly country. Mountains allow him too little real command over his scattered units and he is unable to control them all; in open country, control is a simple matter and does not test his ability to the fullest.

These obvious affinities should guide our planning.

CHAPTER 5

MILITARY VIRTUES OF THE ARMY

MILITARY virtues should not be confused with simple bravery, and still less with enthusiasm for a cause. Bravery is obviously a necessary component. But just as bravery, which is part of the natural make-up of a man's character, can be developed in a soldier—a member of an organization—it must develop differently in him than in other men. In the soldier the natural tendency for unbridled action and out-bursts of violence must be subordinated to demands of a higher kind: obedience, order, rule, and method. An army's efficiency gains life and spirit from enthusiasm for the cause for which it fights, but such enthusiasm is not indispensable.

War is a special activity, different and separate from any other pursued by man. This would still be true no matter how wide its scope, and though every able-bodied man in the nation were under arms. An army's military qualities are based on the individual who is steeped in the spirit and essence of this activity; who trains the capacities it demands, rouses them, and makes them his own; who applies his intelligence to every detail; who gains ease and confidence through practice, and who completely immerses his personality in the appointed task.

No matter how clearly we see the citizen and the soldier in the same man, how strongly we conceive of war as the business of the entire nation, opposed diametrically to the pattern set by the *condottieri* of former times, the business of war will always remain individual and distinct. Consequently for as long as they practise this activity, soldiers will think of themselves as members of a kind of guild, in whose regulations, laws, and customs the spirit of war is given pride of place. And that does seem to be the case. No matter how much one may be inclined to take the most sophisticated view of war, it would be a serious mistake to underrate professional pride (*esprit de corps*) as something that may and must be present in an army to greater or lesser degree. Professional pride is the bond between the various natural forces that activate the military virtues; in the context of this professional pride they crystallize more readily.

An army that maintains its cohesion under the most murderous fire; that cannot be shaken by imaginary fears and resists well-founded ones with all its might; that, proud of its victories, will not lose the strength to obey orders and its respect and trust for its officers even in defeat; whose physical power, like the muscles of an athlete, has been steeled by training in privation and effort; a force that regards such efforts as a means to victory rather than a curse on its cause; that is mindful of all these duties and qualities by virtue of the single powerful idea of the honour of its arms—such an army is imbued with the true military spirit.

It is possible to fight superbly, like the men of the Vendée,* and to achieve great results, like the Swiss, the Americans, and the Spaniards without developing the kind of virtues discussed here; it is even possible to be the victorious commander of a regular army, like Prince Eugène and Marlborough,* without drawing substantially on their help. No one can maintain that it is impossible to fight a successful war without these qualities. We stress this to clarify the concept, and not lose sight of the idea in a fog of generalities and give the impression that military spirit is all that counts in the end. That is not the case. The spirit of an army may be envisaged as a definite moral factor that can be mentally subtracted, whose influence may therefore be estimated—in other words, it is a tool whose power is measurable.

Having thus characterized it, we shall attempt to describe its influence and the various ways of developing it.

Military spirit always stands in the same relation to the parts of an army as does a general's ability to the whole. The general can command only the overall situation and not the separate parts. At the point where the separate parts need guidance, the military spirit must take command. Generals are chosen for their outstanding qualities, and other high-ranking officers are carefully tested; but the testing process becomes less thorough the further we descend on the scale of command, and we must be prepared for a proportionate diminution of personal talent. What is missing here must be made up by military virtues. The same role is played by the natural qualities of a people mobilized for war: *bravery, adaptability, stamina, and enthusiasm*. These, then, are the qualities that can act as substitutes for the military spirit and vice-versa, leading us to the following conclusions:

1. Military virtues are found only in regular armies, and they are

the ones that need them most. In national uprisings and peoples' wars their place is taken by natural warlike qualities, which develop faster under such conditions.

2. A regular army fighting another regular army* can get along without military virtues more easily than when it is opposed by a people in arms; for in the latter case, the forces have to be split up, and the separate units will more frequently have to fend for themselves. Where the troops can remain concentrated, however, the talents of the commander are given greater scope, and can make up for any lack of spirit among the troops. Generally speaking, the need for military virtues becomes greater the more the theatre of operations and other factors tend to complicate the war and disperse the forces.

If there is a lesson to be drawn from these facts, it is that when an army lacks military virtues, every effort should be made to keep operations as simple as possible, or else twice as much attention should be paid to other aspects of the military system. The mere fact that soldiers belong to a 'regular army' does not automatically mean they are equal to their tasks.

Military spirit, then, is one of the most important moral elements in war. Where this element is absent, it must either be replaced by one of the others, such as the commander's superior ability or popular enthusiasm, or else the results will fall short of the efforts expended. How much has been accomplished by this spirit, this sterling quality, this refinement of base ore into precious metal, is demonstrated by the Macedonians under Alexander,* the Roman legions under Caesar,* the Spanish infantry under Alexander Farnese,* the Swedes under Gustavus Adolphus* and Charles XII,* the Prussians under Frederick the Great, and the French under Bonaparte. One would have to be blind to all the evidence of history if one refused to admit that the outstanding successes of these commanders and their greatness in adversity were feasible only with the aid of an army possessing these virtues.

There are only two sources for this spirit, and they must interact in order to create it. The first is a series of victorious wars; the second, frequent exertions of the army to the utmost limits of its strength. Nothing else will show a soldier the full extent of his capacities. The more a general is accustomed to place heavy demands on his soldiers, the more he can depend on their response. A soldier is just as proud of the hardships he has overcome as of the dangers he has faced. In short, the seed will grow only in the soil of constant activity and exertion, warmed by the sun of victory. Once it has grown into a strong tree, it will survive the wildest storms of misfortune and defeat, and even the indolent inertia of peace, at least for a while. Thus, this spirit can be *created* only in war and by great generals, though admittedly it may endure, for several generations at least, even under generals of average ability and through long periods of peace.

One should be careful not to compare this expanded and refined solidarity of a brotherhood of tempered, battle-scarred veterans with the self-esteem and vanity of regular armies which are patched together only by service-regulations and drill. Grim severity and iron discipline may be able to preserve the military virtues of a unit, but it cannot create them. These factors are valuable, but they should not be overrated. Discipline, skill, goodwill, a certain pride, and high morale, are the attributes of an army trained in times of peace. They command respect, but they have no strength of their own. They stand or fall together. One crack, and the whole thing goes, like a glass too quickly cooled. Even the highest morale in the world can, at the first upset, change all too easily into despondency, an almost boastful fear; the French would call it *sauve qui peut*. An army like this will be able to prevail only by virtue of its commander, never on its own. It must be led with more than normal caution until, after a series of victories and exertions, its inner strength will grow to fill its external panoply. We should take care never to confuse the real spirit of an army with its mood.

CHAPTER 11

CONCENTRATION OF FORCES IN SPACE

THE best strategy is always *to be very strong*; first in general, and then at the decisive point. Apart from the effort needed to create military strength, which does not always emanate from the general, there is no higher and simpler law of strategy than that of *keeping one's forces concentrated*. No force should ever be detached from the main body unless the need is definite and *urgent*. We hold fast to this principle, and regard it as a reliable guide. In the course of our analysis, we shall learn in what circumstances dividing one's forces may be justified. We shall also learn that the principle of concentration will not have the same results in every war, but that those will change in accordance with means and ends.

Incredible though it sounds, it is a fact that armies have been divided and separated countless times, without the commander having any clear reason for it, simply because he vaguely felt that this was the way things ought to be done.

This folly can be avoided completely, and a great many unsound reasons for dividing one's forces never be proposed, as soon as concentration of force is recognized as the norm, and every separation and split as an exception that has to be justified.

CHAPTER 13

THE STRATEGIC RESERVE

A RESERVE has two distinct purposes. One is to prolong and renew the action; the second, to counter unforeseen threats. The first purpose presupposes the value of the successive use of force, and therefore does not belong to strategy. The case of a unit being sent to a point that is about to be overrun is clearly an instance of the second category, since the amount of resistance necessary at that point had obviously not been foreseen. A unit that is intended merely to prolong the fighting in a particular engagement and for that purpose is kept in reserve, will be available and subordinate to the commanding officer, though posted out of the reach of fire. Thus it will be a tactical rather than a strategic reserve.

But the need to hold a force in readiness for emergencies may also arise in strategy. Hence there can be such a thing as a strategic reserve, but only when emergencies are conceivable. In a tactical situation, where we frequently do not even know the enemy's measures until we see them, where they may be hidden by every wood and every fold of undulating terrain, we must always be more or less prepared for unforeseen developments, so that positions that turn out to be weak can be reinforced, and so that we can in general adjust our dispositions to the enemy's actions.

Such cases also occur in strategy, since strategy is directly linked to tactical action. In strategy too decisions must often be based on direct observation, on uncertain reports arriving hour by hour and day by day, and finally on the actual outcome of battles. It is thus an essential condition of strategic leadership that forces should be held in reserve according to the degree of strategic uncertainty.

In the defensive generally, particularly in the defence of certain natural features such as rivers, mountain ranges, and so forth, we know this is constantly required.

But uncertainty decreases the greater the distance between strategy and tactics; and it practically disappears in that area of strategy that borders on the political.

The movement of the enemy's columns into battle can be ascertained only by actual observation—the point at which he plans to cross a river by the few preparations he makes, which become apparent a short time in advance; but the direction from which he threatens our country will usually be announced in the press before a single shot is fired. The greater the scale of preparations, the smaller the chance of achieving a surprise. Time and space involved are vast, the circumstances that have set events in motion so well known and so little subject to change, that his decisions will either be apparent early enough, or can be discovered with certainty.

Moreover even if a strategic reserve should exist, in this area of strategy its value will decrease the less specific its intended employment.

We have seen that the outcome of a skirmish or single engagement is in itself of no significance; all such partial actions await resolution in the outcome of the battle as a whole.

In turn, the outcome of the battle as a whole has only relative significance, which varies in numerous gradations according to the size and overall importance of the defeated force. The defeat of a corps may be made up for by the victory of an army, and even the defeat of one army may be balanced or even turned into a victory by the successes of a larger army, as happened in the two days' fighting at Kulm in 1813. No one can doubt this; but it is equally clear that the impact of every victory, the successful outcome of every battle, gains in absolute significance with the importance of the defeated force, and consequently the possibility of recouping such losses at a later encounter also becomes less likely. This point will be examined more closely later on; for the present, it is enough to call attention to the existence of this progression.

Let us add a third observation. While the successive use of force in a tactical situation always postpones the main decision to the end of the action, in strategy the law of the simultaneous use of forces nearly always advances the main decision, which need not necessarily be the ultimate one, to the beginning. These three conclusions, therefore, justify the view that a strategic reserve becomes less essential, less useful, and more dangerous to use, the more *inclusive* and general its intended purpose.

The point at which the concept of a strategic reserve begins to be self-contradictory is not difficult to determine: it comes when the *decisive stage* of the battle has been reached. All forces must be used to achieve it, and any idea of reserves, of *available combat units* that are not meant to be used until after this decision, is an absurdity.

Thus, while a tactical reserve is a means not only of meeting any unforeseen manoeuvre by the enemy but also of reversing the unpredictable outcome of combat when this becomes necessary, strategy must renounce this means, at least so far as the overall decision is concerned. Setbacks in one area can, as a rule, be offset only by achieving gains elsewhere, and in a few cases by transferring troops from one area to another. Never must it occur to a strategist to deal with such a setback by holding forces in reserve.

We have called it an absurdity to maintain a strategic reserve that is not meant to contribute to the overall decision. The point is so obvious that we should not have devoted two chapters to it if it were not for the fact that the idea can look somewhat more plausible when veiled in other concepts, as indeed it frequently is. One man thinks of a strategic reserve as the peak of wise and cautious planning, another rejects the whole idea, including that of a tactical reserve. This kind of confused thinking does actually affect reality. For a striking example, we should recall that in 1806 Prussia billeted a reserve of 20,000 men under Prince Eugene of Württemberg in Brandenburg and could not get them to the Saale River in time, while another 25,000 men were kept in East and south Prussia *to be mobilized at some later stage, to act as a reserve.*

These examples will, we hope, spare us the reproach of tilting at windmills.

CHAPTER 14

ECONOMY OF FORCE*

As we have already said, principles and opinions can seldom reduce the path of reason to a simple line. As in all practical matters, a certain latitude always remains. Beauty cannot be defined by abscissas and ordinates; neither are circles and ellipses created by their algebraic formulas. The man of action must at times trust in the sensitive instinct of judgement, derived from his native intelligence and developed through reflection, which almost unconsciously hits on the right course. At other times he must simplify understanding to its dominant features, which will serve as rules; and sometimes he must support himself with the crutch of established routine.

One of these simplified features, or aids to analysis, is always to make sure that all forces are involved—always to ensure that no part of the whole force is idle. If a segment of one's force is located where it is not sufficiently busy with the enemy, or if troops are on the march—that is, idle—while the enemy is fighting, then these forces are being managed uneconomically. In this sense they are being wasted, which is even worse than using them inappropriately. When the time for action comes, the first requirement should be that all parts must act: even the least appropriate task will occupy some of the enemy's forces and reduce his overall strength, while completely inactive troops are neutralized for the time being. Obviously this view is a corollary of the principles developed in the last three chapters.* It is the same truth, restated from a somewhat broader point of view, and reduced to a single concept.

CHAPTER 16

THE SUSPENSION OF ACTION IN WAR

IF we regard war as an act of mutual destruction, we are bound to think of both sides as usually being in action and advancing. But as soon as we consider each moment separately, we are almost equally bound to think of only one side as advancing while the other is expectantly waiting; for conditions will never be exactly identical on both sides, nor will their mutual relationship remain the same. In time changes will occur, and it follows that any given moment will favour one side more than the other. If we assume that both generals are completely cognizant of their own and their opponent's conditions, one of them will be motivated to act, which becomes in turn to the other a reason for waiting. Both cannot simultaneously want to advance, or on the other hand to wait. This mutual exclusion of identical aims does not, in the present context, derive from the principle of polarity, and therefore it does not contradict the assertion made in Chapter Five of Book Two.* Rather, its basis lies in the fact that the determinant is really the same for both commanders: the probability of improvement, or deterioration, of the situation in the future.

Even if we suppose that circumstances could be completely balanced, or if we assume that insufficient knowledge of their mutual circumstances gives the commanders the impression that such equality exists, the differences in their political purpose will still rule out the possibility of a standstill. Politically, only one can be the aggressor: there can be no war if both parties seek to defend themselves. The aggressor has a positive aim, while the defender's aim is merely negative. Positive action is therefore proper to the former, since it is the only means by which he can achieve his ends. Consequently when conditions are equal for both parties the attacker ought to act, since his is the positive aim.

Seen in this light, suspension of action in war is a contradiction in terms. Like two incompatible elements, armies must continually destroy one another. Like fire and water they never find themselves in a state of equilibrium, but must keep on interacting until one of them has completely disappeared. Imagine a pair of wrestlers deadlocked and inert for hours on end! In other words, military action ought to run its course steadily like a wound-up clock. But no matter how savage the nature of war, it is fettered by human weaknesses; and no one will be surprised at the contradiction that man seeks and creates the very danger that he fears.

The history of warfare so often shows us the very opposite of unceasing progress toward the goal, that it becomes apparent that *immobility* and *inactivity* are the normal *state* of armies in war, and *action is the exception.* This might almost make us doubt the accuracy of our argument. But if this is the burden of much of military history, the most recent series of wars does substantiate the argument. Its validity was demonstrated and its necessity was proved only too plainly by the revolutionary wars.* In these wars, and even more in the campaigns of Bonaparte, warfare attained the unlimited degree of energy that we consider to be its elementary law. We see it is possible to reach this degree of energy; and if it is possible, it is necessary.

How, in fact, could we reasonably defend the exertion of so much effort in war, unless action is intended! A baker fires his oven only when he is ready to bake bread; horses are harnessed to a carriage only when we intend to drive; why should we make the enormous exertions inherent in war if our only object is to produce a similar effort on the part of the enemy?

So much in justification of the general principle. Now for its modifications, insofar as they arise from the nature of the subject and do not depend on individual circumstances.

Let us note three determinants that function as inherent counterweights and prevent the clockwork from running down rapidly or without interruption.

The first of these, which creates a permanent tendency toward delay and thus becomes a retarding influence, is the fear and indecision native to the human mind. It is a sort of moral force of gravity, which, however, works by repulsion rather than attraction: namely, aversion to danger and responsibility

In the fiery climate of war, ordinary natures tend to move more ponderously; stronger and more frequent stimuli are therefore needed to ensure that momentum is maintained. To understand why the war is being fought is seldom sufficient in itself to overcome this ponderousness. Unless an enterprising martial spirit is in command, a man who is as much at home in war as a fish is in water, or unless great responsibilities exert a pressure, inactivity will be the rule, and progress the exception.

The second cause is the imperfection of human perception and judgement, which is more pronounced in war than anywhere else. We hardly know accurately our own situation at any particular moment, while the enemy's, which is concealed from us, must be deduced from very little evidence. Consequently it often happens that both sides see an advantage in the same objective, even though in fact it is more in the interest of only one of them. Each may therefore think it wiser to await a better moment, as I have already explained in Chapter Five of Book Two.*

The third determinant, which acts like a ratchet-wheel, occasionally stopping the works completely, is the greater strength of the defensive. A may not feel strong enough to attack B, which does not, however, mean that B is strong enough to attack A. The additional strength of the defensive is not only lost when the offensive is assumed but is transferred to the opponent. Expressed in algebraic terms, the difference between A + B and A − B equals 2 B. It therefore happens that both sides at the same time not only feel too weak for an offensive, but that they really are too weak.

Thus, in the midst of the conflict itself, concern, prudence, and fear of excessive risks find reason to assert themselves and to tame the elemental fury of war.

But these determinants are hardly adequate explanations for the long periods of inactivity that occurred in earlier wars, in which no vital issues were at stake, and in which nine-tenths of the time that the troops spent under arms was occupied by idleness. As stated in the chapter on the Purpose and Means in War, this phenomenon is mainly due to the influence that the demands of the one belligerent, and the condition and state of mind of the other, exert on the conduct of the war.

These factors can become so influential that they reduce war to something tame and half-hearted. War often is nothing more than armed neutrality, a threatening attitude meant to support negotiations, a mild attempt to gain some small advantage before sitting back and letting matters take their course, or a disagreeable obligation imposed by an alliance, to be discharged with as little effort as possible.

In all such cases, where the impetus of interest is slight and where there is little hostile spirit, where we neither want to do much harm to the enemy nor have much to fear from him, in short where no great motive presses and promotes action, governments will not want to risk much. This explains the tame conduct of such conflicts, in which the hostile spirit of true war is held in check.

The more these factors turn war into something half-hearted, the less solid are the bases that are available to theory: essentials become rarer, and accidents multiply.

Nevertheless, even this type of conflict gives scope to intelligence; possibly even wider and more varied scope. Gambling for high stakes seems to have turned into haggling for small change. In this type of war, where military action is reduced to insignificant, time-killing flourishes, to skirmishes that are half in earnest and half in jest; to lengthy orders that add up to nothing; to positions and marches that in retrospect are described as scientific, simply because their minute original motive has been forgotten and common sense cannot make anything of them—in this type of conflict many theorists see the real, authentic art of war. In these feints, parries, and short lunges of earlier wars they find the true end of all theory and the triumph of mind over matter. More recent wars appear to them as crude brawls that can teach nothing and that are to be considered as relapses into barbarism. This view is as petty as its subject. In the absence of great forces and passions it is indeed simpler for ingenuity to function; but is not guiding great forces, navigation through storms and surging waves, a higher exercise of the intellect? That other, formalized type of swordsmanship is surely included and implicit in the more energetic mode of conducting war. It has the same relation to it as the movements on a ship have to the motion of the ship. It can only be carried on so long as it is tacitly understood that the opponent follows suit. But is it possible to tell how long this condition will be observed? The French Revolution surprised us in the false security of our ancient skills, and drove us from Châlons to Moscow. With equal suddenness, Frederick the Great surprised the Austrians in the quiet of their antiquated ways of war, and shook their monarchy to its foundations. Woe to the government, which, relying on half-hearted politics and a shackled military policy, meets a foe who, like the untamed elements, knows no law other than his own power! Any defect of action and effort will turn to the advantage of the enemy, and it will not be easy to change from a fencer's position to that of a wrestler. A slight blow may then often be enough to cause a total collapse.

All of these reasons explain why action in war is not continuous but spasmodic. Violent clashes are interrupted by periods of observation, during which both sides are on the defensive. But usually one side is more strongly motivated, which tends to affect its behaviour: the offensive element will dominate, and usually maintain its continuity of action.

BOOK SIX
DEFENCE

CHAPTER I

ATTACK AND DEFENCE

1. The Concept of Defence

WHAT is the concept of defence? The parrying of a blow. What is its characteristic feature? Awaiting the blow. It is this feature that turns any action into a defensive one; it is the only test by which defence can be distinguished from attack in war. Pure defence, however, would be completely contrary to the idea of war, since it would mean that only one side was waging it Therefore, defence in war can only be relative, and the characteristic feature of waiting should be applied only to the basic concept, not to all of its components. A partial engagement is defensive if we await the advance, the charge of the enemy. A battle is defensive if we await the attack—await, that is, the appearance of the enemy in front of our lines and within range. A campaign is defensive if we wait for our theatre of operations to be invaded. In each of these cases the characteristic of waiting and parrying is germane to the general idea without being in conflict with the concept of war; for we may find it advantageous to await the charge against our bayonets and the attack on our position and theatre of operations. But if we are really waging war, we must return the enemy's blows; and these offensive acts in a defensive war come under the heading of 'defence'—in other words, our offensive takes place within our own positions or theatre of operations. Thus, a defensive campaign can be fought with offensive battles, and in a defensive battle, we can employ our divisions offensively. Even in a defensive position awaiting the enemy assault, our bullets take the offensive. So the defensive form of war is not a simple shield, but a shield made up of well-directed blows.

2. Advantages of Defence

What is the object of defence? Preservation. It is easier to hold ground than take it. It follows that defence is easier than attack, assuming both sides have equal means. Just what is it that makes preservation and protection so much easier? It is the fact that time which is allowed to pass unused accumulates to the credit of the defender. He reaps where he did not sow. Any omission of attack—whether from bad judgement, fear, or indolence—accrues to the defenders' benefit. This saved Prussia from disaster more than once during the Seven Years War.* It is a benefit rooted in the concept and object of defence: it is in the nature of all defensive action. In daily life, and especially in litigation (which so closely resembles war) it is summed up by the Latin proverb *beati sunt possidentes.* Another benefit, one that arises solely from the nature of war, derives from the advantage of position, which tends to favour the defence.

Having outlined these general concepts, we now turn to the substance.

Tactically, every engagement, large or small, is defensive if we leave the initiative to our opponent and await his appearance before our lines. From that moment on we can employ all offensive means without losing the advantages of the defensive—that is to say the advantages of waiting and the advantages of position. At the strategic level the campaign replaces the engagement and the theatre of operations takes the place of the position. At the next stage, the war as a whole replaces the campaign, and the whole country the theatre of operations. In both cases, defence remains the same as at the tactical level.

We have already indicated in general terms that defence is easier than attack. But defence has a passive purpose: *preservation*; and attack a positive one: *conquest.* The latter increases one's own capacity to wage war; the former does not. So in order to state the relationship precisely, we must say that *the defensive form of warfare is intrinsically stronger than the offensive.* This is the point we have been trying to make, for although it is implicit in the nature of the matter and experience has confirmed it again and again, it is at odds with prevalent opinion, which proves how ideas can be confused by superficial writers.

If defence is the stronger form of war, yet has a negative object, it follows that it should be used only so long as weakness compels, and be abandoned as soon as we are strong enough to pursue a positive object. When one has used defensive measures successfully, a more favourable balance of strength is usually created; thus, the natural course in war is to begin defensively and end by attacking. It would therefore contradict the very idea of war to regard defence as its final purpose, just as it would to regard the passive nature of defence not only as inherent in the whole but also in all its parts. In other words, a war in which victories were used only defensively without the intention of counterattacking would be as absurd as a battle in which the principle of absolute defence—passivity, that is—were to dictate every action.

The soundness of this general idea could be challenged by citing many examples of wars in which the ultimate purpose of defence was purely defensive, without any thought being given to a counter-offensive. This line of argument would be possible if one forgot that a general concept is under discussion. The examples that could be cited to prove the opposite must all be classed as cases in which the possibility of a counteroffensive had not yet arisen.

In the Seven Years War, for instance, Frederick the Great had no thought of taking the offensive, at least not in its final three years. Indeed, we believe that in this war he always regarded offensives solely as a better means of defence. This attitude was dictated by the general situation; and it is natural for a commander to concentrate only on his immediate needs. Nevertheless one cannot look at this example of defence on a grand scale without speculating that the idea of a possible counteroffensive against Austria may have been at the root of it, and conclude that the time for such a move had not yet come. The peace that was concluded proves that this was not an empty assumption: What else could have induced the Austrians to make peace but the thought that their forces could not on their own outweigh the genius of the King; that in any case they would have to increase their efforts; and that any relaxation was almost bound to cost them further territory? And, indeed, is there any doubt that Frederick would have tried to crush the Austrians in Bohemia and Moravia again if Russia, Sweden, and the Army of the Empire had not diverted his energies?

Now that we have defined the concept of defence and have indicated its limits, we return once more to our claim that defence is *the stronger form of waging war.*

Close analysis and comparison of attack and defence will prove the point beyond all doubt. For the present, we shall merely indicate the inconsistencies the opposite view involves when tested by experience. If attack were the stronger form, there would be no case for using the defensive, since its purpose is only passive. No one would want to do anything but attack: defence would be pointless.

Conversely, it is natural that the greater object is bought by greater sacrifice. Anyone who believes himself strong enough to employ the weaker form, attack, can have the higher aim in mind; the lower aim can only be chosen by those who need to take advantage of the stronger form, defence. Experience shows that, given two theatres of operations, it is practically unknown for the weaker army to attack and the stronger stay on the defensive. The opposite has always happened everywhere, and amply proves that commanders accept defence as the stronger form, even when they personally would rather attack.

Some related points remain to be discussed in the following chapters.

CHAPTER 3

THE RELATIONSHIP BETWEEN ATTACK AND DEFENCE IN STRATEGY

LET us again begin by examining the factors that assure strategic success.

As we have said before, in strategy there is no such thing as victory. Part of strategic success lies in timely preparation for a tactical victory; the greater the strategic success, the greater the likelihood of a victorious engagement. The rest of strategic success lies in the exploitation of a victory won. The more strategy has been able, through its ingenuity, to exploit a victorious battle; the more that it can wrest out of the collapsing edifice whose foundations have been shattered by the action; the more completely the fruits of the hard-won victory can be harvested; then the greater the success. The main factors responsible for bringing about or facilitating such a success—thus the main factors in strategic effectiveness—are the following:

1. The advantage of terrain
2. Surprise—either by actual assault or by deploying unexpected strength at certain points
3. Concentric attack (all three as in tactics)

4. Strengthening the theatre of operations, by fortresses, with all they involve
5. Popular support
6. The exploitation of moral factors.[1]

What is the relationship of attack and defence with regard to these factors?

In strategy as well as in tactics, the defence enjoys the advantage of terrain, while the attack has the advantage of initiative. As regards surprise and initiative, however, it must be noted that they are infinitely more important and effective in strategy than in tactics. Tactical initiative can rarely be expanded into a major victory, but a strategic one has often brought the whole war to an end at a stroke. On the other hand, the use of this device assumes *major, decisive, and exceptional* mistakes on the enemy's part. Consequently it will not do much to tip the scales in favour of attack.

Surprising the enemy by concentrating superior strength at certain points is again comparable to the analogous case in tactics. If the defender were compelled to spread his forces over several points of access, the attacker would obviously reap the advantage of being able to throw his full strength against any one of them.

Here too the new system of defence has, by its new approach, imperceptibly introduced new principles. Where the defender has no reason to fear that his opponent will be able by advancing along an undefended road to seize an important depot or munitions dump, or take a fortress unawares, or even the capital unawares; where, therefore, he is not forced to attack the enemy on the road chosen by the latter in order to avoid having his retreat cut off; then there is no reason for him to split his forces. If the attacker chooses a road on which he does not expect to meet the defender, the latter can still seek him out there with his entire strength a few days later. Indeed he can be sure that in most cases the attacker himself will oblige him by seeking him out. But if for some reason the attacker has to advance with divided forces—and problems of supply often leave

---

[1] Anyone who has learned his strategy from Herr von Bülow will not understand how it is that we have simply left out the whole of Bülow's teaching. But it is not our fault if Bülow deals with minor matters only. An office boy would be just as puzzled if he searched the index of an arithmetic book and found no entry for such practical rules as the rules of three or five. But Herr von Bülow's opinions can hardly be counted as practical rules. We have made the comparison for other reasons.

him little choice—the defender obviously reaps the benefit of being able to attack a part of his opponent with his own full strength.

In strategy, the nature of flank and rear attacks on a theatre of operations changes to a significant degree.

1. The effect of cross fire is eliminated, since one cannot fire from one end of a theatre of operations to the other.
2. There is less fear of being cut off, since whole areas cannot be sealed off in strategy as they can in tactics.
3. Because of the greater areas involved in strategy, the effectiveness of interior and therefore shorter lines is accentuated and forms an important counterbalance against concentric attacks.
4. A new factor emerges in the vulnerability of lines of communication, that is, in the consequences of their being cut.

Because of the greater areas involved in strategy, envelopment or concentric attack will of course only be possible for the side which takes the initiative—in other words, the attacker. The defender cannot, as he can in tactics, surround the surrounder in turn, for he cannot deploy his troops in the relative depth required, nor keep them sufficiently concealed. But what use to the attack is ease of envelopment if its rewards do not materialize? In strategy, therefore, there would be no justification at all in putting forward the enveloping attack as a means of victory, were it not for its effect on lines of communication. Yet this is seldom an important factor at the earliest stage when attack is first confronted by defence, and the two sides face each other in their opening positions. It only begins to tell in the course of a campaign, when the attacker, in enemy territory, gradually becomes the defender. At that point the new defender finds his lines of communication weakening, and the original defender can exploit that weakness once he has taken the offensive. But it must be obvious that as a rule the defender deserves no credit for this advantage, since it really derives from the principles inherent in the defence itself.

The fourth element, *the advantages of the theatre of operations*, naturally benefit the defender. By initiating the campaign, the attacking army cuts itself off from its own theatre of operations, and suffers by having to leave its fortresses and depots behind. The larger the area of operations that it must traverse, the more it is weakened—by the effect of marches and by the detachment of garrisons. The defending army, on the other hand, remains intact. It benefits from its fortresses, nothing depletes its strength, and it is closer to its sources of supply.

The support of the population, the fifth principle, will not necessarily apply to every defence; a defensive campaign may be fought in enemy territory. Still, this element derives from the concept of defence alone, and it is applicable in the vast majority of cases. What is meant is primarily (but not exclusively) the effectiveness of militia, and arming the population. Furthermore, every kind of friction is reduced, and every source of supply is nearer and more abundant.

The campaign of 1812* will here serve as a magnifying glass, for it clearly reveals how the third and fourth of these factors can operate. Half a million men crossed the Niemen; only 120,000 fought at Borodino, and still fewer reached Moscow.

One may say indeed that the outcome of this enormous effort was so great that even if the Russians had not followed it up with their own counter-offensive, they would have been secure from any fresh invasion for a long time to come. Of course no European country, except for Sweden, is in a similar position to Russia's; but the principle is universal and differs only in degree.

As to the fourth and fifth factors, one might add that these assets pertain to the basic case of defence in one's own country. If defence is moved to enemy soil and gets involved in offensive operations, it will be transformed into a further liability of the offensive, in much the same way as with the third element mentioned above. The offensive is not composed of active elements alone, any more than the defensive is made up solely of passive elements. Indeed, any attack that does not immediately lead to peace must end on the defensive.

Thus, if all elements of defence that occur during an offensive are weakened by the very fact that they are part of the offensive, then we must regard this as another general liability pertaining to it.

This is not simply hairsplitting. Far from it: this is the greatest disadvantage of all offensive action. Hence when a strategic attack is being planned one should from the start give very close attention to this point—namely, the defensive that will follow. The matter will be discussed in greater detail in the book on strategic planning.

The important moral forces that sometimes permeate war like a leaven may occasionally be used by a commander to invigorate his troops. These forces may be found on the side of defence as well as that of attack; at least one can say that the ones which especially favour attack, such as panic and confusion in the enemy's ranks, do not normally emerge until after the decisive blow has been struck, and so seldom have much bearing on its course.

All this should suffice to justify our proposition that *defence is a stronger form of war than attack*. But we still have to mention a minor factor that so far has been left out of account. It is courage: the army's sense of superiority that springs from the awareness that one is taking the initiative. This affinity is a real one, but it is soon overlaid by the stronger and more general spirit that an army derives from its victories or defeats, and by the talent or incompetence of its commander.

CHAPTER 5

THE CHARACTER OF STRATEGIC DEFENCE

WE have already stated what defence is—simply the more effective form of war: a means to win a victory that enables one to take the offensive after superiority has been gained; that is, to proceed to the active object of the war.

Even when the only point of the war is to maintain the *status quo*, the fact remains that merely parrying a blow goes against the essential nature of war, which certainly does not consist merely in enduring. Once the defender has gained an important advantage, defence as such has done its work. While he is enjoying this advantage, he must strike back, or he will court destruction. Prudence bids him strike while the iron is hot and use the advantage to prevent a second onslaught. How, when, and where that reaction is to begin depends, of course, on many other conditions which we shall detail subsequently. For the moment we shall simply say that this transition to the counterattack must be accepted as a tendency inherent in defence—indeed, as one of its essential features. Wherever a victory achieved by the defensive form is not turned to military account, where, so to speak, it is allowed to wither away unused, a serious mistake is being made.

A sudden powerful transition to the offensive—the flashing sword of vengeance—is the greatest moment for the defence. If it is not in the commander's mind from the start, or rather if it is not an integral part of his idea of defence, he will never be persuaded of the superiority of the defensive form; all he will see is how much of the enemy's resources he can destroy or capture. But these things do not depend on the way in which the knot is tied, but on the way in which it is untied. Moreover, it is a crude error to equate attack with the idea of assault alone, and therefore, to conceive of defence as merely misery and confusion.

Admittedly, an aggressor often decides on war before the innocent defender does, and if he contrives to keep his preparations sufficiently secret, he may well take his victim unawares. Yet such surprise has nothing to do with war itself, and should not be possible. War serves the purpose of the defence more than that of the aggressor. It is only aggression that calls forth defence, and war along with it. The aggressor is always peace-loving (as Bonaparte always claimed to be); he would prefer to take over our country unopposed. To prevent his doing so one must be willing to make war and be prepared for it. In other words it is the weak, those likely to need defence, who should always be armed in order not to be overwhelmed. Thus decrees the art of war.

When one side takes the field before the other, it is usually for reasons that have nothing to do with the intention of attack or defence. They are not the motives, but frequently the result of an early appearance. The side that is ready first and sees a significant advantage in a surprise attack, will for *that* reason take the offensive. The side that is slower to prepare can to some degree make up for the consequent disadvantage by exploiting the advantages of defence.

Generally speaking, however, the ability to profit from being the first to be ready must be considered an advantage to the attacker, as we have acknowledged in Book Three. Still, this general advantage is not essential in every specific case.

Consequently, if we are to conceive of defence as it should be, it is this. All means are prepared to the utmost; the army is fit for war and familiar with it; the general will let the enemy come on, not from confused indecision and fear, but by his own choice, coolly and deliberately; fortresses are undaunted by the prospect of a siege; and finally a stout-hearted populace is no more afraid of the enemy than

he of it. Thus constituted, defence will no longer cut so sorry a
figure when compared to attack, and the latter will no longer look so
easy and infallible as it does in the gloomy imagination of those who
see courage, determination, and movement in attack alone, and in
defence only impotence and paralysis.

CHAPTER 7

INTERACTION BETWEEN ATTACK
AND DEFENCE

THE time has come to consider defence and attack separately, insofar
as they can be separated. We shall start with defence for the following reasons. While it is quite natural and even indispensable to base
the principles of defence on those that govern attack and vice versa,
there must be a third aspect to one of them that serves as a point of
departure for the whole chain of ideas and makes it tangible. Our
first question, therefore, concerns this point.

Consider in the abstract how war originates. Essentially, the concept of war does not originate with the attack, because the ultimate
object of attack is not fighting: rather, it is possession. The idea of
war originates with the defence, which does have fighting as its
immediate object, since fighting and parrying obviously amount to
the same thing. Repulse is directed only toward an attack, which is
therefore a prerequisite to it; the attack, however, is not directed
toward defence but toward a different goal—possession, which is not
necessarily a prerequisite for war. It is thus in the nature of the case
that the side that first introduces the element of war, whose point of
view brings two parties into existence, is also the side that establishes
the initial laws of war. That side is the *defence*. What is under discussion here is not a specific instance but a general, abstract case, which
must be postulated to advance theory.

We now know where to find the fixed point that is located outside
the interaction of attack and defence: it lies with the defence.

If this argument is correct, the defender must establish ground
rules for his conduct even if he has no idea what the attacker means
to do, and these ground rules must certainly include the disposition

of his forces. The attacker, on the other hand, so long as he knows nothing about his adversary, will have no guidelines on which to base the use of his forces. All he can do is to take his forces with him—in other words, take possession by means of his army. Indeed, that is what actually happens: for it is one thing to assemble an army and another to use it. An aggressor may take his army with him on the chance that he may have to use it, and though he may take possession of a country by means of his army instead of officials, functionaries, and proclamations, he has not yet, strictly speaking, committed a positive act of war. It is the defender, who not only concentrates his forces but disposes them in readiness for action, who first commits an act that really fits the concept of war.

We now come to the second question: what in theory is the nature of the underlying causes that initially motivate the defence, before it has even considered the possibility of being attacked? Obviously, it is an enemy's advance with a view to taking possession, which we have treated as extraneous to war but which forms the basis for the initial steps of military activity. This advance is meant to deter defence, and it must, therefore, be thought of in relation to the country; and this is what produces the initial general dispositions of the defence. Once these have been established, the attack will be directed toward them, and new ground rules of defence will be based on an examination of the means used by the attack. At this point the interaction has become evident, and theorists may continue to study it as long as new results appear and make the study seem worthwhile.

This brief analysis was necessary to provide somewhat greater clarity and substance to our subsequent discussion; it is not intended for the battlefield, nor for any future general, but for the legions of theorists who, up to now, have treated such questions far too lightly.

CHAPTER 8

TYPES OF RESISTANCE

THE essence of defence lies in parrying the attack. This in turn implies waiting, which for us is the main feature of defence and also its chief advantage.

Since defence in war cannot simply consist of passive endurance, waiting will not be absolute either, but only relative. In terms of space, it relates to the country, the theatre of operations, or the position; in terms of time, to the war, the campaign, or the battle. True these are not unalterable units, but the central points of certain areas that overlap and merge with one another. In practice, however, one must often be satisfied with merely arranging things into categories rather than strictly separating them; and those terms, in general usage, have become clearly enough defined to serve as nuclei around which other ideas may conveniently be gathered.

The defender of a country, therefore, merely awaits the attack on his country, the defender of a theatre of war awaits the attack on that theatre, and the defender of a position awaits the attack on that position. Once the enemy has attacked, any active and therefore more or less offensive move made by the defender does not invalidate the concept of defence, for its salient feature and chief advantage, *waiting*, has been established.

The concepts characteristic of time—war, campaign and battle— are parallel to those of space—country, theatre of operations and position—and so bear the same relation to our subject.

Defence is thus composed of two distinct parts, waiting and acting. By linking the former to a definite object that precedes action, we have been able to merge the two into one whole. But a defensive action—especially a large-scale one such as a campaign or a war— will not, in terms of time, consist of two great phases, the first of which is pure waiting and the second pure action; it will alternate between these two conditions, so that waiting may run like a continuous thread through the whole period of defence.

The nature of the matter demands that so much importance should be attached to waiting. To be sure, earlier theorists never gave it the status of an independent concept, but in practice it has continuously served as a guideline, though for the most part men were not consciously aware of it. Waiting is such a fundamental feature of all warfare that war is hardly conceivable without it, and hence we shall often have occasion to revert to it by pointing out its effect in the dynamic play of forces.

We should now like to elucidate how the principle of waiting runs through the entire period of defence, and how the successive stages of defence originate in it.

In order to establish our ideas by means of a simpler example, we shall defer (till we reach the book on war plans) the defence of a country, a more diversified subject, and one that is more strongly influenced by political circumstances. On the other hand, defence in a position or in a battle is a tactical matter; only when it is *completed* can it serve as the starting point of strategic activity. Therefore, we shall take the defence of a *theatre of operations* as the subject that will best illustrate the conditions of defence.

We have pointed out that waiting and acting—the latter always being a riposte and therefore a reaction—are both essential parts of defence. Without the former, it would not be defence, without the latter, it would not be war. This conception has already led us to argue that *defence is simply the stronger form of war, the one that makes the enemy's defeat more certain.* We must insist on this interpretation—partly because any other will eventually lead to absurdity, partly because the more vivid and total this impression, the more it will strengthen the total act of defence.

It would be contrary to this interpretation to discuss reaction, the second necessary component of defence, by making a distinction between its parts, and considering that phase which, strictly speaking, consists in warding off the enemy—from the country, the theatre of operations, the position—as the only *necessary* part, which would be limited to what is needed to achieve those purposes. The other phase, the possibility of a reaction that *expands into the realm of actual strategic offence*, would then have to be considered as being foreign to, and unconnected with, defence. Such a distinction is basically unacceptable: we must insist that the idea of *retaliation* is fundamental to all defence. Otherwise, no matter how much damage the first phase of reaction, if successful, may have done to the enemy, the proper balance would still be wanting in the restoration of the dynamic relationship between attack and defence.

We repeat then that defence is the stronger form of war, the one that makes the enemy's defeat more certain. It may be left to circumstances whether or not a victory so gained exceeds the original purpose of the defence.

Since defence is tied to the idea of waiting, the aim of defeating the enemy will be valid only on the condition that there is an attack. If no attack is forthcoming, it is understood that the defence will be content to hold its own; so this is its aim, or rather its primary aim, during the period of waiting. The defence will be able to reap the benefits of the stronger form of war only if it is willing to be satisfied with this more modest goal.

Let us postulate an army that has been ordered to defend its theatre of operations. It may do this in the following ways:

1. It can attack the enemy the moment he invades its theatre of operations (Mollwitz, Hohenfriedberg).*

2. It can take up position near the frontier, wait until the enemy appears and is about to attack, and then attack him first (Czaslau, Soor, Rossbach).* Such an attitude is obviously more passive; it calls for a longer period of waiting; and though little or no *time* may be gained by the second plan as compared to the first if the enemy really does attack, still, the battle that was certain in the first case will be less certain in the second, and it may turn out that the enemy's determination will not extend as far as an attack. The advantage of waiting, therefore, has become greater.

3. It can wait, not merely for the enemy's decision to attack—that is, his appearance in full view of the position—but also for the actual attack (as at Bunzelwitz,* to take another example from the campaigns of the commander we have been referring to).* In that event the army will fight a true defensive battle, but one, as we have said before, that may include offensive moves by some part of the army. Here too, as in the previous case, the gain in time is immaterial, but the enemy's determination will be tested once again. Many an army has advanced to the attack but refrained at the last moment, or desisted after the first attempt on finding the enemy's position too strong.

4. It can withdraw to the interior of the country and resist there. The purpose of this withdrawal is to weaken the attacker to such an extent that one can wait for him to break off his advance of his own accord, or, at least, be too weak to overcome the resistance with which he will eventually be confronted.

The simplest and most outstanding example would be the case in which the defender is able to leave one or more fortresses behind, which the attacker must invest or besiege. It is obvious that this will weaken his forces and provide an opportunity for an attack by the defender at a point where he has the upper hand.

Even where there are no fortresses, such a retreat to the interior can gradually restore to the defender the balance or superiority that he did not have on the frontier. In a strategic attack, every advance reduces the attacker's strength, partly as an absolute loss and partly because of the division of forces which becomes necessary. We shall discuss this in greater detail in connection with the attack. For the present, we shall simply assume this statement to be correct, since it has been sufficiently demonstrated in past wars.

The main advantage of this fourth case lies in the time that is gained. If the enemy lays siege to our fortresses, we have gained time until their surrender (which is probable, but which may take several weeks, and in some cases months). If, on the other hand, his loss of strength, the exhaustion of the momentum of his attack, is caused simply by his advance and by having to leave garrisons at vital points, and thus only by the distance he has covered, the amount of time gained will usually be even greater and we are not so strongly compelled to act at any given moment.

Not only will the relative strength of defender and attacker have changed when this action has run its course, the former will also have to his credit the increased benefit of waiting. Even if the attacker has not been weakened enough by his advance to prevent him from attacking our main force where it has come to rest, he may lack the determination to do so. This determination must be stronger here than it would have had to be at the frontier: the reason is partly that his forces are reduced and no longer fresh while his danger has increased, and partly that irresolute commanders will completely forget about the necessity of a battle once possession of the area has been achieved; either because they really think it is no longer necessary, or because they are glad of the pretext. Their failure to attack is not, of course, the adequate negative success for the defender that it would have been on the frontier, but the time gained is substantial nonetheless.

In all four cases cited, it goes without saying that the defender has the benefit of terrain, and that the support of his fortresses and the populace are favourable to his action. With each successive stage of defence these elements become more significant, and in the fourth stage they are particularly effective in weakening the enemy. Since the advantages of waiting also increase with each phase, it follows that each successive stage of defence is more effective than the last, and that this form of warfare gains in effectiveness the further it is removed from attack. We are not afraid of being accused on this account of believing that the most passive kind of defence is the strongest. Each successive stage, far from being intended to weaken the act of resistance, is meant merely to *prolong and postpone* it. Surely there is no contradiction in saying that one is able to resist more effectively in a strong and suitably entrenched position, and that, after the enemy has wasted half his strength on it, a counter-attack will be that much more effective. Daun could hardly have won at Kolin* without his strong position. If his pursuit of the mere 18,000 men whom Frederick was able to lead from the field had been more energetic, this victory could have been one of the most brilliant in the annals of war.

What we do maintain is that with each successive stage of defence the defender's predominance or, more accurately, his counterweight will increase, and so in consequence will the strength of his reaction.

Can we say that the advantages that derive from an intensified defence are to be had without cost? Not at all: the sacrifices with which they must be purchased will increase equally.

Whenever we wait for the enemy inside our own theatre of operations, no matter how close to the frontier the decisive action may be fought, the enemy's forces will enter our theatre of operations, which will entail sacrifices in this area. If we had attacked him first, the damage would have been incurred by him. The sacrifices tend to increase whenever we fail to advance toward the enemy in order to attack him; the area he occupies and the time he takes to advance to our position will continue to increase them. If we intend to give defensive battle and thus leave the initiative and the timing up to the enemy, the possibility exists that he may well remain for a considerable time in the area he holds. So the time we gain by his postponement of the decision has to be paid for in this manner. The sacrifices become even more noticeable in the case of a retreat into the interior of the country.

However, the reduction of the defender's strength that is caused by all of these sacrifices will usually affect his fighting forces only later, not immediately: it is frequently so indirect as to be barely noticeable. Thus the defender tries to increase his immediate strength by paying for it later—in other words, he borrows like anyone else who needs more than he has.

In order to assess the results of these various forms of resistance, we have to examine *the purpose of the enemy's attack*. It is to gain possession of our theatre of operations, or at least a substantial part of it; for the concept of the whole implies at least the greater part, and a strip a few miles wide is seldom of independent strategic importance. Therefore, so long as the attacker is not in possession, so long, in other words, as fear of our strength has prevented him from entering our theatre of operations or seeking out our position or has caused him to avoid the battle we are prepared to give, the objects of the defence have been accomplished. Our defensive dispositions have proved successful. Admittedly, this is only a negative success which will not directly produce enough strength for a real counter-attack. But it may do so in an indirect way, gradually: the time that passes is *lost to the aggressor*. Time lost is always a disadvantage that is bound in some way to weaken him who loses it.

Thus, in the first three stages of defence (in other words, those taking place at the border) *the very lack of a decision constitutes a success for the defence.*

That, however, is not the case in the fourth stage.

If the enemy lays siege to our fortresses we must relieve them in good time—in other words, it is up to us to bring about a decision by positive action.

This is also the case where the enemy has pursued us into the interior without besieging any of our fortresses. While we may have more time and can wait until the enemy is at his weakest, the assumption will remain that we shall have to take the initiative in the end. Indeed, the enemy may by then have taken all of the area that was the object of his attack, but he holds it as a loan. The tension continues to exist, and the decision is still to come. So long as the defender's strength increases every day while the attacker's diminishes, the absence of a decision is in the former's best interest; but if only because the effects of the general losses to which the defender has continually exposed himself are finally catching up with him, the point of culmination will necessarily be reached when the defender must make up his mind and act, when the advantages of waiting have been completely exhausted.

There is of course no infallible means of telling when that point has come; a great many conditions and circumstances may determine it. We should note, however, that the approach of winter is usually the most natural turning point. If we cannot prevent the enemy from wintering in the area he has occupied, we might as well give it up for lost. Even so there is the example of Torres Vedras to remind us that this is not a universal rule.

What is it that, broadly speaking, constitutes a decision?

In our discussion, we have always assumed decision to occur in the form of battle, but that is not necessarily so. We can think of any number of engagements by smaller forces that may lead to a change in fortune, either because they really end in bloodshed, or because the probability of their consequences necessitate the enemy's retreat.

No other kind of decision is possible in the theatre of operations itself: that necessarily follows from the concept of war we have proposed. Even where the hostile army is forced to retreat because of lack of food, this factor does after all arise out of the limitations that our forces impose. If our army were not present, the enemy would surely find ways of helping himself.

So even when his offensive has run its course, when the enemy has become the victim of the difficult conditions of the advance and has been weakened and reduced by hunger, by sickness and the need to detach troops, it is really the fear of our fighting forces alone that makes him turn about and abandon all he has gained. Nevertheless there is a vast difference between such a decision and one that has been reached at the border.

At the border, his arms are faced by our arms—they alone hold him in check or do him damage. But when his offensive has run its course, he has been worn out largely by his own efforts. This will impart a completely different value to our arms, which are no longer the only factor in the decision though they may be the ultimate one. The ground has been prepared for it by the breakdown of the enemy's forces during their advance to the extent where a retreat, and a complete reversal of the situation, can be caused by the mere possibility of our reaction. In such a case one is bound to be realistic, to give credit for the decision to the difficulties of the offensive. Admittedly, one will not be able to find an example in which the defending forces were not also a factor, but for the sake of practical considerations, it is important to distinguish which of these two factors was dominant.

In the light of these ideas we think it is fair to say that two decisions, and therefore two kinds of reaction, are possible on the defending side, depending on whether the attacker is to *perish by the sword* or *by his own exertions*.

It is obvious that the first type of decision will predominate in the first three stages of defence, and the second type in the fourth. Indeed, the latter can essentially only take place where the retreat penetrates deeply into the interior of the country. It is in fact the only reason that can justify such a retreat and the great sacrifices it entails.

Two basically different types of resistance have now been identified. There are cases in military history where they stand out as clearly and distinctly as any abstract concept ever can in practice. In 1745 Frederick the Great attacked the Austrians at Hohenfriedberg as they descended from the Silesian mountains—at a time when their strength could not have been sapped either by exertions or by the detachment of troops. Wellington,* on the other hand, stayed in the fortified lines of Torres Vedras* until cold and hunger had left Masséna's army* so depleted that it withdrew of its own accord. In that case the defender's forces took no part in the actual process of wearing down the enemy. At other times where both types are closely linked, one will still be distinctly dominant. Take, for instance, 1812. In that famous campaign so many savage engagements were fought that, under different circumstances, an absolute decision might well have been reached by the sword alone; yet there is probably no other case in which the evidence is so clear that the invader was destroyed by his own exertions. Only about 90,000 of the 300,000 men who made up the French centre reached Moscow. Only 13,000 men had been detached. Casualties, therefore, numbered 197,000, of which fighting certainly cannot account for more than a third.

All campaigns that are known for their so-called temporizing, like those of the famous Fabius Cunctator,* were calculated primarily to destroy the enemy by making him exhaust himself.

In general there have been many campaigns that were won on that principle without anyone explicitly saying so. One will arrive at the true cause of many decisions only by ignoring the far-fetched explanations of historians, and instead closely examining the events themselves.

We believe that we have thus adequately described the considerations that underlie defence and its various phases. By pointing out these two chief means of resistance, we hope we have explained clearly how the principle of waiting runs through the whole system and combines with the principle of positive action in such a way that the latter may appear early in one case and late in another. After which, the advantages of waiting will be seen to be exhausted.

We believe that we have now surveyed as well as delimited the whole field of defence. True, there still are aspects that are important enough to deserve a chapter on their own, points on which a series of reflections could be based and which ought not to be overlooked: for instance, the nature and influence of fortresses and entrenched camps, the defence of mountains and rivers, flanking operations, and so forth. We shall deal with these in the chapters that follow. Still, none of these subjects seems to fall outside the scope of the ideas explained above, but merely constitute their application to specific places and circumstances. The foregoing sequence of ideas has developed out of the concept of defence and its relation to attack. We have linked those simple ideas with reality, and so demonstrated how to move from reality to these simple ideas and achieve a solid analytic base. In the course of debate we will therefore not need to resort to arguments that themselves are ephemeral.

But armed resistance, by its diversity of possible combinations, can so change the appearance and vary the character of armed defence, especially in cases where there is no actual fighting but the outcome is affected by the fact that there could be, that one is almost tempted to think some new effective principle here awaits discovery. The vast difference between savage repulse in a straightforward battle, and the effect of a strategic web that prevents things from getting that far, will lead one to assume that a different force must be at work—a conjecture somewhat like that of the astronomers' who deduced from the enormous void between Mars and Jupiter that other planets must exist.

If an attacker finds the enemy in a strong position that he thinks he cannot take, or on the far side of a river that he believes to be impassable, or even if he fears he will jeopardize his food supply by advancing any further, it is still only the force of the defender's arms which produces these results. What actually halts the aggressor's action is the fear of defeat by the defender's forces, either in major engagements or at particularly important points; but he is not likely to concede this, at least not openly.

One may admit that even where the decision has been bloodless, it was determined in the last analysis by engagements that did not take place but *had merely been offered*. In that case, it will be argued, *the strategic planning* of these engagements, rather than the tactical decision, should be considered the operative principle. Moreover such strategic planning would be dominant only in cases where defence is conducted by some means other than force of arms. We admit this; but it brings us to the very point we wanted to make. What we say in fact is this: where the tactical results of the engagement are assumed to be the *basis* of all strategic plans, it is always possible, and a serious risk, that the attacker will proceed on that basis. He will endeavour above all to be tactically superior, in order to upset the enemy's strategic planning The latter, therefore, can never be considered *as something independent*: it can only become valid when one has reason to be confident of tactical success. To illustrate briefly what we mean, let us recall that a general such as Bonaparte could ruthlessly cut through all his enemies' strategic plans in search of battle, because he seldom doubted the battle's outcome. So whenever the strategists did not endeavour with all their might to crush him in battle with superior force, whenever they engaged in subtler (and weaker) machinations, their schemes were swept away like cobwebs. Schemes of that sort would have been enough to check a general like Daun; but it would have been folly to oppose Bonaparte and his army in the way the Prussians handled Daun and the Austrians in the Seven Years War. Why? Because Bonaparte was well aware that everything turned on tactical results, and because he could rely on them, while Daun's situation was different in both respects. *That is why* we think it is useful to emphasize that all strategic planning rests on tactical success alone, and that—whether the solution is arrived at in battle or not—this is in all cases the actual fundamental basis for the decision. Only when one has no need to fear the outcome— because of the enemy's character or situation or because the two armies are evenly matched physically and psychologically or indeed because one's own side is the stronger—only then can one expect results from strategic combinations *alone*.

When we look at the history of war and find a large number of campaigns in which the attacker broke off his offensive without having fought a decisive battle, consequently where strategic combinations appear effective, we might believe that such combinations have at least great inherent power, and that they would normally decide the outcome on their own whenever one did not need to assume a decisive superiority of the offensive in tactical situations. Our answer here must be that this assumption, too, is erroneous in situations that arise in the theatre of operations and are therefore part of war itself. The reason for the ineffectiveness of most attacks lies in the general, the political conditions of war.

The general conditions from which a war arises, and that form its natural basis, will also determine its character. This will later be discussed in greater detail, under the heading of war plans. But these general conditions have transformed most wars into mongrel affairs, in which the original hostilities have to twist and turn among conflicting interests to such a degree that they emerge very much attenuated. This is bound to affect the offensive, *the side of positive action*, with particular strength. It is not surprising, therefore, that one can stop such a breathless, hectic attack by the mere flick of a finger. Where resolution is so faint and paralysed by a multitude of considerations that it has almost ceased to exist, a mere show of resistance will often suffice.

We can see that in many cases the reason for the defender's being successful without having to fight does not lie in the fact that he occupies many impregnable positions, nor in the size of the mountain ranges that lie across the theatre of operations, nor in the broad stream that traverses it, nor in the ease with which the threatened blow can be paralysed by a well-planned series of engagements. The real reason is the faintness of the attacker's determination, which makes him hesitate and fear to move.

Countervailing forces of this kind can and must be reckoned with; but they must be recognized for what they are, rather than having their effects attributed to other causes—those that alone concern us here. We must state emphatically that, in this respect, military history can well become a chronic lie and deception if critics fail to apply the required correctives.

At this point let us examine, in their most common form, the vast number of offensive campaigns that failed without a decisive battle being fought.

The aggressor marches into hostile territory; he drives the enemy back a little, but then begins to have doubts about risking a decisive battle. He halts and faces his opponent, acting as if he had made a conquest and was interested only in protecting it—in short, he behaves as if it were the enemy's affair to seek a battle, as if he himself were ready to fight at any time, and so forth. All of these are mere *pretexts*, which a general uses to delude his army, his government, the world at large, and even himself. The truth of the matter is that the enemy's position has been found too strong. Here we are not talking of a case in which the aggressor fails to attack because a victory would be of no use to him, because his advance having run its course he does not have enough resiliency to start a new one. This would assume that a successful attack had already taken place and resulted in a genuine conquest; rather, we have in mind a case in which the aggressor gets bogged down in the middle of an intended conquest.

At that point the attacker will wait for a favourable turn of events to exploit. There is as a rule no reason to expect such a favourable turn: the very fact that an attack had been intended implies that the immediate future promises no more than the present. It is therefore a fresh delusion. If, as is usual, the operation is a joint one timed to coincide with others, the other armies will then be blamed for his failures. By way of excusing his inaction he will plead inadequate support and cooperation. He will talk of insuperable obstacles, and look for motives in the most intricately complicated circumstances. So he will fritter his strength away in doing nothing, or rather in doing too little to bring about anything but failure. Meanwhile the defender is gaining time—which is what he needs most. The season is getting late, and the whole offensive ends with the return of the invader to his winter quarters in his own theatre of operations.

This tissue of falsehoods ends by passing into history in place of the obvious and simple truth: that failure was due to *fear of the enemy's forces*. When the critics begin to study a campaign of this sort they tend to get lost in argument and counterargument. No convincing answer will be found, because everything is guesswork and the critics never dig deep enough to find the truth.

That sort of fraudulence is not merely a matter of bad habit; its roots lie in the nature of the case. The counterweights that weaken the elemental force of war, and particularly the attack, are primarily located in the political relations and intentions of the government, which are concealed from the rest of the world, the people at home, the army, and in some cases even from the commander. For instance no one can and will admit that his decision to stop or to give up was motivated by the fear that his strength would run out, or that he might make new enemies or that his own allies might become too strong. That sort of thing is long kept confidential, possibly forever. Meanwhile, a plausible account must be circulated. The general is, therefore, urged, either for his own sake or the sake of his government, to spread a web of lies. This constantly recurring shadow-boxing in the dialectics of war has, as theory, hardened into systems, which are, of course, equally misleading. Only a theory that will follow the simple thread of internal cohesion as we have tried to make ours do, can get back to the essence of things.

If military history is read with this kind of scepticism, a vast amount of verbiage concerning attack and defence will collapse, and the simple conceptualization we have offered will automatically emerge. We believe that it is valid for the whole field of defence, and that only if we cling to it firmly can the welter of events be clearly understood and mastered.

Let us now examine the employment of these various methods of defence.

They are all intensifications of the same thing, each one exacting increased sacrifices on the part of the defender. A general's choice, all other things being equal, would largely be determined by this fact. He would choose the method he considered adequate to give his forces the necessary degree of resistance; but to avoid unnecessary losses he would not retreat any further. It must be admitted, however, that the choice of different methods is already severely limited by other major factors that play a part in defence and are bound to urge him to use one method or another. A withdrawal to the interior calls for ample space, or else it requires conditions like those that obtained in Portugal* in 1810: one ally (England) provided solid support to the rear, while another (Spain), with its extensive territory, reduced a great deal of the enemy's impact. The location of fortresses—close to the border or farther inland—may also decide for or against a certain method. Even more decisive are the nature of the terrain and the country, and the character, customs, and temper of its people. The choice between an offensive and defensive battle may be determined by the enemy's plans or by the characteristics of both armies and their generals. Finally, the possession or lack of an outstanding position or defensive line may lead to one method or the other. In short, a mere listing of these factors is enough to indicate that in defence they are more influential on the choice than is relative strength alone. Since we shall become more familiar with the most important factors that have here only been touched upon, we will later be able to demonstrate in greater detail what influence they exert on the choice. Finally, the implications will be treated in a comprehensive analysis in the book on war plans and campaign plans.

That influence, however, will normally become decisive only if the relative strengths are not too disproportionate. Where they are (and therefore in the majority of cases), relative strength will prevail. The history of war is full of proof that this has actually occurred — quite apart from the chain of reasoning developed here — *through the hidden processes of intuitive judgement*, like almost everything that happens in war. It was the same general, with the same army, who, in the same theatre of operations, fought the battle of Hohenfriedberg and also moved into camp at Bunzelwitz. Thus, even Frederick the Great, who when it came to a battle was the most offensive-minded of generals, was finally compelled to resort to a strict defensive when the disproportion of strength became too great. Indeed did not Bonaparte, who used to rush at his enemies like a wild boar, twist and turn like a caged animal when the ratio of forces was no longer in his favour in August and September 1813, without attempting a reckless attack on any one of his enemies? And do we not find him at Leipzig,* in October of the same year, when the disparity of forces had reached its peak, taking refuge in the angle made by the Parthe, Elster, and Pleisse rivers, as if he were cornered in a room with his back to the wall, waiting for his enemies?

We should like to add that this chapter, more than any other of our work, shows that our aim is not to provide new principles and methods of conducting war; rather, we are concerned with examining the essential content of what has long existed, and to trace it back to its basic elements.

CHAPTER 26

THE PEOPLE IN ARMS

IN the civilized parts of Europe, war by means of popular uprisings is a phenomenon of the nineteenth century.* It has its advocates and its opponents. The latter object to it either on political grounds, considering it as a means of revolution, a state of legalized anarchy that is as much of a threat to the social order at home as it is to the enemy; or else on military grounds, because they feel that the results are not commensurate with the energies that have been expended.

The first objection does not concern us at all: here we consider a general insurrection as simply another means of war—in its relation, therefore, to the enemy. The second objection, on the other hand, leads us to remark that a popular uprising should, in general, be considered as an outgrowth of the way in which the conventional barriers have been swept away in our lifetime by the elemental violence of war. It is, in fact, a broadening and intensification of the fermentation process known as war. The system of requisitioning, and the enormous growth of armies resulting from it and from universal conscription, the employment of militia—all of these run in the same direction when viewed from the standpoint of the older, narrower military system, and that also leads to the calling out of the home guard and arming the people.

The innovations first mentioned were the natural, inevitable consequences of the breaking down of barriers. They added so immensely to the strength of the side that first employed them that the opponent was carried along and had to follow suit. That will also hold true of the people's war. Any nation that uses it intelligently will, as a rule, gain some superiority over those who disdain its use. If this is so, the question only remains whether mankind at large will gain by this further expansion of the element of war; a question to which the answer should be the same as to the question of war itself. We shall leave both to the philosophers. But it can be argued that the resources expended in an insurrection might be put to better use in other kinds of warfare. No lengthy investigation is needed, however, to uncover the fact that these resources are, for the most part, not otherwise available and cannot be disposed of at will. Indeed, a significant part of them, the psychological element, is called into being only by this type of usage.

When a whole nation renders armed resistance, the question then is no longer, 'Of what value is this to the people,' but 'what is its potential value, what are the conditions that it requires, and how is it to be utilized.'

By its very nature, such scattered resistance will not lend itself to major actions, closely compressed in time and space. Its effect is like that of the process of evaporation: it depends on how much surface is exposed. The greater the surface and the area of contact between it and the enemy forces, the thinner the latter have to be spread, the greater the effect of a general uprising. Like smouldering embers, it consumes the basic foundations of the enemy forces. Since it needs time to be effective, a state of tension will develop while the two elements interact. This tension will either gradually relax, if the insurgency is suppressed in some places and slowly burns itself out in others, or else it will build up to a crisis: a general conflagration closes in on the enemy, driving him out of the country before he is faced with total destruction. For an uprising by itself to produce such a crisis presupposes an occupied area of a size that, in Europe, does not exist outside Russia, or a disproportion between the invading army and the size of the country that would never occur in practice. To be realistic, one must therefore think of a general insurrection within the framework of a war conducted by the regular army, and coordinated in one all-encompassing plan.

The following are the only conditions under which a general uprising can be effective:

1. The war must be fought in the interior of the country.
2. It must not be decided by a single stroke.
3. The theatre of operations must be fairly large.
4. The national character must be suited to that type of war.
5. The country must be rough and inaccessible, because of mountains, or forests, marshes, or the local methods of cultivation.

The relative density of the population does not play a decisive part; rarely are there not enough people for the purpose. Nor does it make much difference whether the population is rich or poor—at least it should not be a major consideration, although one must remember that poor men, used to hard, strenuous work and privation, are generally more vigorous and more warlike.

One peculiarity of the countryside that greatly enhances the effectiveness of an insurrection is the scattered distribution of houses and farms, which, for instance, can be found in many parts of Germany. Under such conditions the country will be more cut up and thickly wooded, the roads poorer if more numerous; the billeting of troops will prove infinitely more difficult, and, above all, the most characteristic feature of insurgency in general will be constantly repeated in miniature: the element of resistance will exist everywhere and nowhere. Where the population is concentrated in villages, the most restless communities can be garrisoned, or even looted and burned down as punishment; but that could scarcely be done in, say, a Westphalian farming area.

Militia and bands of armed civilians cannot and should not be employed against the main enemy force—or indeed against any sizeable enemy force. They are not supposed to pulverize the core but to nibble at the shell and around the edges. They are meant to operate in areas just outside the theatre of war—where the invader will not appear in strength—in order to deny him these areas altogether. Thunder clouds of this type should build up all around the invader the farther he advances. The people who have not yet been conquered by the enemy will be the most eager to arm against him; they will set an example that will gradually be followed by their neighbours. The flames will spread like a brush fire, until they reach the area on which the enemy is based, threatening his lines of communication and his very existence. One need not hold an exaggerated faith in the power of a general uprising, nor consider it as an inexhaustible, unconquerable force, which an army cannot hope to stop any more than man can command the wind or the rain—in short, one need not base one's judgement on patriotic broadsides in order to admit that peasants in arms will not let themselves be swept along like a platoon of soldiers. The latter will cling together like a herd of cattle and generally follow their noses; peasants, on the other hand, will scatter and vanish in all directions, without requiring a special plan. This explains the highly dangerous character that a march through mountains, forests, or other types of difficult country can assume for a small detachment: at any moment the march may turn into a fight. An area may have long since been cleared of enemy troops, but a band of peasants that was long since driven off by the head of a column may at any moment reappear at its tail. When it comes to making roads unusable and blocking narrow passes, the means available to outposts or military raiding parties and those of an insurgent peasantry have about as much in common as the movements of an automaton have with those of a man. The enemy's only answer to militia actions is the sending out of frequent escorts as protection for his convoys, and as guards on all his stopping places, bridges, defiles, and the rest. The early efforts of the militia may be fairly weak, and so will these first detachments, because of the dangers of dispersal. But the flames of insurrection will be fanned by these small detachments, which will on occasion be overpowered by sheer numbers; courage and the appetite for fighting will rise, and so will the tension, until it reaches the climax that decides the outcome.

A general uprising, as we see it, should be nebulous and elusive; its resistance should never materialize as a concrete body, otherwise the enemy can direct sufficient force at its core, crush it, and take many prisoners. When that happens, the people will lose heart and, believing that the issue has been decided and further efforts would be useless, drop their weapons. On the other hand, there must be some concentration at certain points: the fog must thicken and form a dark and menacing cloud out of which a bolt of lightning may strike at any time. These points for concentration will, as we have said, lie mainly on the flanks of the enemy's theatre of operations. That is where insurgents should build up larger units, better organized, with parties of regulars that will make them look like a proper army and enable them to tackle larger operations. From these areas the strength of the insurgency must increase as it nears the enemy's rear, where he is vulnerable to its strongest blows. The larger groups are intended to harass the more considerable units that the enemy sends back; they will also arouse uneasiness and fear, and deepen the psychological effect of the insurrection as a whole. Without them the impression would not be sufficiently great, nor would the general situation give the enemy enough cause for alarm.

A commander can more easily shape and direct the popular insurrection by supporting the insurgents with small units of the regular army. Without these regular troops to provide encouragement, the local inhabitants will usually lack the confidence and initiative to take to arms. The stronger the units detailed for the task, the greater their power of attraction and the bigger the ultimate avalanche. But there are limiting factors. For one thing, it could be fatal to the army to be frittered away on secondary objectives of that kind—to be dissolved, so to speak, in the insurgency—merely to form a long and tenuous defensive line, which is a sure way of destroying army and insurgents alike. For another, experience tends to show that too many regulars in an area are liable to decimate the vigour and effectiveness of a popular uprising by attracting too many enemy troops; also, the inhabitants will place too much reliance upon the regulars; and finally, the presence of considerable numbers of troops taxes the local resources in other ways, such as billets, transportation, requisitions, and so forth.

Another means of avoiding an effective enemy reaction to a popular uprising is, at the same time, one of the basic principles of insurrection: it is the principle of seldom, or never, allowing this important strategic means of defence to turn into tactical defence. *Insurgent actions* are similar in character to all others fought by second-rate troops: they start out full of vigour and enthusiasm, but there is little level-headedness and tenacity in the long run. Moreover, not much is lost if a body of insurgents is defeated and dispersed—that is what it is for. But it should not be allowed to go to pieces through too many men being killed, wounded or taken prisoner: such defeats would soon dampen its ardour. Both these characteristics are entirely alien to the nature of a tactical defence. A defensive action ought to be a slow, persistent, calculated business, entailing a definite risk; mere attempts that can be broken off at will can never lead to a successful defence. So if the defence of a sector is entrusted to the home guard, one must avoid getting involved in a major defensive battle, or else they will perish no matter how favourable the circumstances. They may and should defend the points of access to a mountain area or the dikes across a marsh or points at which a river can be crossed for as long as possible; but once these are breached, they had better scatter and continue their resistance by means of surprise attacks, rather than huddle together in a narrow redoubt, locked into a regular defensive position from which there is no escape. No matter how brave a people is, how warlike its traditions, how great its hatred for the enemy, how favourable the ground on which it fights: the fact remains that a national uprising cannot maintain itself where the atmosphere is too full of danger. Therefore, if its fuel is to be fanned into a major conflagration, it must be at some distance, where there is enough air, and the uprising cannot be smothered by a single stroke.

This discussion has been less an objective analysis than a groping for the truth. The reason is that this sort of warfare is not as yet very common; those who have been able to observe it for any length of time have not reported enough about it. We merely wish to add that strategic plans for defence can provide for a general insurrection in one of two ways: either as a last resort after a defeat or as a natural auxiliary before a decisive battle. The latter use presupposes a withdrawal to the interior and the form of indirect defence described in Chapters Eight and Twenty-Four of this book. Therefore, we shall add only a few words concerning the calling out of the home guard after a battle has been lost.

A government must never assume that its country's fate, its whole existence, hangs on the outcome of a single battle, no matter how decisive. Even after a defeat, there is always the possibility that a turn of fortune can be brought about by developing new sources of internal strength or through the natural decimation all offensives suffer in the long run or by means of help from abroad. There will always be time enough to die; like a drowning man who will clutch instinctively at a straw, it is the natural law of the moral world that a nation that finds itself on the brink of an abyss will try to save itself by any means.

No matter how small and weak a state may be in comparison with its enemy, it must not forgo these last efforts, or one would conclude that its soul is dead. The possibility of avoiding total ruin by paying a high price for peace should not be ruled out, but even this intention will not, in turn, eliminate the usefulness of new measures of defence. They will not make the peace more difficult and onerous, but easier and better. They are even more desirable where help can be expected from other states that have an interest in our survival. A government that after having lost a major battle, is only interested in letting its people go back to sleep in peace as soon as possible, and, overwhelmed by feelings of failure and disappointment, lacks the courage and desire to put forth a final effort, is, because of its weakness, involved in a major inconsistency in any case. It shows that it did not deserve to win, and, possibly for that very reason was unable to.

With the retreat of the army into the interior—no matter how complete the defeat of a state—the potential of fortresses and general insurrections must be evoked. In this respect, it will be advantageous if the flanks of the main theatre of operations are bordered by mountains or other difficult terrain, which will then emerge as bastions, raking the invader with their strategic enfilade.

Once the victor is engaged in sieges, once he has left strong garrisons all along the way to form his line of communication, or has even sent out detachments to secure his freedom of movement and keep adjoining provinces from giving him trouble; once he has been weakened by a variety of losses in men and matériel, the time has come for the defending army to take the field again. Then a well-placed blow on the attacker in his difficult situation will be enough to shake him.

BOOK SEVEN
THE ATTACK

CHAPTER I

ATTACK IN RELATION TO DEFENCE

WHERE two ideas form a true logical antithesis, each complementary to the other, then fundamentally each is implied in the other. If the limitations of our mind do not allow us to comprehend both simultaneously, and discover by antithesis the whole of one in the whole of the other, each will nevertheless shed enough light on the other to clarify many of its details. In consequence we believe that the earlier chapters about defence will have sufficiently illuminated the aspects of attack on which they touch. But this is not always so. No analytical system can ever be explored exhaustively. It is natural that, where the antithesis does not lie so close to the root of the concept as in the previous chapters, what we can say about attack will not follow directly from what was said there about defence. A shift in our viewpoint will bring us nearer the subject, so that we can examine more closely what we previously surveyed from a distance. This will supplement our previous analysis; and what will now be said about attack will frequently also cast more light on defence.

In dealing with attack, we shall largely have to treat topics that have already been discussed. But we do not think we need proceed, as do so many textbooks in engineering, by circumventing or demolishing all the positive values we identified in defence and proving that for every method of defence there is an infallible method of attack. Defence has its strengths and weaknesses. Though the former may not be insurmountable, the cost of surmounting them may be disproportionate. This must hold true whatever way we look at it; otherwise we are contradicting ourselves. Nor do we intend to analyse this interaction exhaustively. Every method of defence leads to a method of attack, but this is often so obvious that we do not need to discuss both in order to perceive it: one follows automatically from the other. We intend to indicate in each case the special features of attack that do not arise directly from the defence. This is bound to call for a number of chapters that have no counterpart in the previous book.

CHAPTER 2

THE NATURE OF STRATEGIC ATTACK

As we have seen, defence in general (including of course strategic defence) is not an absolute state of waiting and repulse; it is not total, but only relative passive endurance. Consequently, it is permeated with more or less pronounced elements of the offensive. In the same way, the attack is not a homogeneous whole: it is perpetually combined with defence. The difference between the two is that one cannot think of the defence without that necessary component of the concept, the counterattack. This does not apply to the attack. The offensive thrust or action is complete in itself. It does not have to be complemented by defence; but dominating considerations of time and space do introduce defence as a necessary evil. *In the first place*, an attack cannot be completed in a single steady movement: periods of rest are needed, during which the attack is neutralized, and defence takes over automatically. *Second*, the area left in rear of the advancing forces, an area vital to their existence, is not necessarily covered by the attack, and needs special protection.

The act of attack, particularly in strategy, is thus a constant alternation and combination of attack and defence. The latter, however, should not be regarded as a useful preliminary to the attack or an intensification of it, and so an active principle; rather it is simply a necessary evil, an impeding burden created by the sheer weight of the mass. It is its original sin, its mortal disease.

We call it an *impeding* burden: unless defence contributes to the attack, it will tend to diminish its effect, if only because of the loss of time involved. Is it possible for this defensive component, which is part of every offensive, to be actually disadvantageous? When we assume *attack to be the weaker and defence the stronger form* of war, it seems to follow that the latter cannot be detrimental to the former: if there are enough forces to serve the weaker form, they must surely suffice for the stronger. That is generally so. We shall examine the subject more closely in the chapter on the *culminating point of victory*. However, we must not forget that the superiority of *strategic defence* arises partly from the fact that the attack itself cannot exist without some measure of defence—and defence of a much less effective kind. What was true of defence as a whole no longer holds true for these parts, and it thus becomes clear how these features of defence may positively weaken the attack. It is these very moments of weak defence during an offensive that the positive activity of the offensive principle *in defence* seeks to exploit.

Consider the difference of the situations during the twelve-hour rest period that customarily follows a day's action. The defender holds a well-chosen position which he knows and has prepared with care; the attacker has stumbled into his bivouac like a blind man. A longer halt, such as may be required to obtain supplies, await reinforcements, and so forth, will find the defender close to his fortresses and depots, while the attacker is like a bird perched on a limb. Every attack will anyhow end in a defence whose nature will be decided by the circumstances. These may be very favourable when the enemy forces have been destroyed, but where this is not the case things may be very difficult. Even though this type of defence is no longer part of the offensive, it must affect it and help determine its effectiveness.

It follows that every attack has to take into account the defence that is necessarily inherent in it, in order clearly to understand its disadvantages and to anticipate them.

But in other respects attack remains consistent and unchanged, while defence has its stages, insofar as the principle of waiting is exploited. From these, essentially different forms of action will result, as has been discussed in the chapter on kinds of resistance.

But since attack has but one single active principle (defence in this case being merely a dead weight that clings to it) one will find in it no such differentiations. Admittedly there are tremendous differences in terms of vigour, speed, and striking power, but these are differences of degree, not of kind. It even might be conceivable for the attacker to choose the defensive form to further his aims. He might, for instance, occupy a strong position in the hope that the defender would attack him there. But such cases are so rare that in the light of actual practice they do not require consideration in our listing of concepts and principles. To sum up: there is no growth of intensity in an attack comparable to that of the various types of defence.

Finally, the means of attack available are usually limited to the fighting forces—to which one must of course add any fortresses located close to the theatre of war, which may have a substantial influence on the attack. But this influence will weaken as the advance proceeds; clearly, the attacker's fortresses can never play so promin- ent a part as the defender's, which often become a main feature. Popular support of the attack is conceivable where the inhabitants are more favourably inclined toward the attacker than toward their own army. Finally, the attacker may have allies, but only as a result of special or fortuitous circumstances. Their support is not inherent in the nature of the attack. Thus, while we have included fortresses, popular uprisings and allies among the possible means of defence, we cannot include them among the means of attack. In the first they are inherent, in the second they are rare and then usually accidental.

CHAPTER 3

THE OBJECT OF STRATEGIC ATTACK

IN war, the subjugation of the enemy is the end, and the destruction of his fighting forces the means. That applies to attack and defence alike. By means of the destruction of the enemy's forces defence leads to attack, which in turn leads to the conquest of the country. That, then, is the objective, but it need not be the whole country; it may be limited to a part—a province, a strip of territory, a fortress, and so forth. Any one of these may be of political value in negotiations, whether they are retained or exchanged.

The object of strategic attack, therefore, may be thought of in numerous gradations, from the conquest of a whole country to that of an insignificant hamlet. As soon as the objective has been attained the attack ends and the defence takes over. One might therefore think of a strategic attack as an entity with well-defined limits. But practice—seeing things, that is, in the light of actual events—does not bear this out. In practice the stages of the offensive—that is, the intentions and the actions taken—as often turn into defensive action as defensive plans grow into the offensive. It is rare, or at any rate uncommon, for a general to set out with a firm objective in mind; rather, he will make it dependent on the course of events. Frequently his attack may lead him further than he expected; after a more or less brief period of rest he often acquires new strength; but this should not be considered as a second, wholly separate action. At other times he may be stopped earlier than he had anticipated, but without abandoning his plan and moving over to a genuine defensive. So it becomes clear that if a successful defence can imperceptibly turn into attack, the same can happen in reverse. These gradations must be kept in mind if we wish to avoid a misapplication of our general statements on the subject of attack.

CHAPTER 4

THE DIMINISHING FORCE OF THE ATTACK

THE diminishing force of the attack is one of the strategist's main concerns. His awareness of it will determine the accuracy of his estimate in each case of the options open to him.

Overall strength is depleted:

1. If the object of the attack is to occupy the enemy's country (Occupation normally begins only after the first decisive action, but the attack does not cease with this action.)
2. By the invading armies' need to occupy the area in their rear so as to secure their lines of communication and exploit its resources
3. By losses incurred in action and through sickness
4. By the distance from the source of replacements
5. By sieges and the investment of fortresses
6. By a relaxation of effort
7. By the defection of allies.

But these difficulties may be balanced by other factors that tend to strengthen the attack. Yet it is clear that the overall result will be determined only after these various quantities have been evaluated. For instance, a weakening of the attack may be partially or completely cancelled out or outweighed by a weakening of the defence. This is unusual; in any case one should never compare all the forces in the field, but only those facing each other at the front or at decisive points. Different examples: the French in Austria and Prussia, and in Russia; the allies in France; the French in Spain.*

CHAPTER 5

THE CULMINATING POINT OF THE ATTACK

SUCCESS in attack results from the availability of superior strength, including of course both physical and moral. In the preceding chapter we pointed out how the force of an attack gradually diminishes; it is possible in the course of the attack for superiority to increase, but usually it will be reduced. The attacker is purchasing advantages that may become valuable at the peace table, but he must pay for them on the spot with his fighting forces. If the superior strength of the attack—which diminishes day by day—leads to peace, the object will have been attained. There are strategic attacks that have led directly to peace, but these are the minority. Most of them only lead up to the point where their remaining strength is just enough to maintain a defence and wait for peace. Beyond that point the scale turns and the reaction follows with a force that is usually much stronger than that of the original attack. This is what we mean by the culminating point of the attack. Since the object of the attack is the possession of the enemy's territory, it follows that the advance will continue until the attacker's superiority is exhausted; it is this that drives the offensive on toward its goal and can easily drive it further. If we remember how many factors contribute to an equation of forces, we will understand how difficult it is in some cases to determine which side has the upper hand. Often it is entirely a matter of the imagination.

What matters therefore is to detect the culminating point with discriminative judgement. We here come up against an apparent contradiction. If defence is more effective than attack, one would think that the latter could never lead too far; if the less effective form is strong enough the more effective form should be even stronger.*

CHAPTER 6

DESTRUCTION OF THE ENEMY'S FORCES

DESTRUCTION of the enemy's forces is the means to the end. What does this mean? At what price?

Different points of view that are possible:

1. To destroy only what is needed to achieve the object of the attack
2. To destroy as much as possible
3. The preservation of one's own fighting forces as the dominant consideration
4. This can go so far that the attacker will attempt destructive action only under favourable circumstances, which may also apply to the achievement of the objective, as has been mentioned in Chapter Three.

The engagement is the only means of destroying the enemy's forces, but it may act in two different ways, either directly or indirectly, by a combination of engagements. Thus while a battle is the principal means, it is not the only one. The capture of a fortress or a strip of territory also amounts to a destruction of enemy forces. It may lead to further destruction, and thereby become an indirect means as well.

So the occupation of an undefended strip of territory may, aside from its direct value in achieving an aim, also have value in terms of destruction of enemy forces. Manoeuvring the enemy out of an area he has occupied is not very different from this, and should be considered in the same light, rather than as a true success of arms. These means are generally overrated; they seldom achieve so much as a battle, and involve the risk of drawbacks that may have been overlooked. They are tempting because they cost so little.

They should always be looked upon as minor investments that can only yield minor dividends, appropriate to limited circumstances and weaker motives. But they are obviously preferable to pointless battles—victories that cannot be fully exploited.

CHAPTER 7

THE OFFENSIVE BATTLE

WHAT we have said about the defensive battle will have already cast considerable light on the offensive battle.

We were thinking of the kind of battle in which the defensive is most prominent, in order to clarify the nature of the defensive. But very few battles are of that type; most of them are in part encounters (*demi-rencontres*) in which the defensive element tends to get lost. This is not so with the offensive battle, which retains its character under all circumstances, and can assert it all the more since the defender is not in his proper element. So there remains a certain difference in the character of the battle—the way in which it is conducted by one side or the other—between those battles that are not really defensive and those that are true encounters (*rencontres*). The main feature of an offensive battle is the outflanking or by-passing of the defender—that is, taking the initiative.

Enveloping actions obviously possess great advantages; they are, however, a matter of tactics. The attacker should not forgo these advantages simply because the defender has a means of countering them; it is a means the attacker cannot use, for it is too much bound up with the rest of the defender's situation. A defender, in order to outflank an enemy who is trying to outflank him, must be operating from a well-chosen, well-prepared position. Even more important is the fact that the defender cannot actually use the full potential offered by his situation. In most cases, defence is a sorry, makeshift affair; the defender is usually in a tight and dangerous spot in which, because he expects the worst, he meets the attack half-way. Consequently, battles that make use of enveloping lines or reversed fronts—which ought to be the result of advantageous lines of communication—tend in reality to be the result of moral and physical superiority. For examples, see Marengo, Austerliz, and Jena.* And in the opening battle of a campaign the attacker's base-line, even if it is not superior to the defender's, will usually be very wide, because the frontier is so close, and he can thus afford to take risks. Incidentally a flank-attack—that is, a battle in which the front has been shifted—is more effective than an enveloping one. It is a mistake to assume that an enveloping strategic advance must be linked with it from the start, as it was at Prague.* They seldom have anything in common, and the latter is a very precarious business about which we shall have more to say when we discuss the attack on a theatre of operations. Just as the commander's aim in a defensive battle is to postpone the decision as long as possible in order to gain time (because a defensive battle that remains undecided at sunset can usually be considered a battle won), the aim of the commander in an offensive battle is to expedite the decision. Too much haste, on the other hand, leads to the risk of wasting one's forces. A peculiarity in most offensive battles is doubt about the enemy's position; they are characterized by groping in the dark—as, for example, at Austerlitz, Wagram, Hohenlinden,* Jena, and Katzbach.* The more this is so, the more it becomes necessary to concentrate one's forces, and to outflank rather than envelop the enemy. In Chapter Twelve of Book Four* it has been demonstrated that the real fruits of victory are won only in pursuit. By its very nature, pursuit tends to be a more integral part of the action in an offensive battle than in a defensive one.

CHAPTER 15

ATTACK ON A THEATRE OF WAR: SEEKING A DECISION

MOST aspects of this question have already been touched upon in Book Six, 'On Defence', which will have reflected sufficient light on the subject of attack.

The concept of a self-contained theatre of operations is in any case more closely associated with defence than with attack. A number of salient points, such as *the object of the attack* and *the sphere of effectiveness of the victory*, have already been dealt with in Book Six, and the really basic and essential features of attack can be expounded only in connection with the subject of war plans. Still, enough remains to be set forth here, and we shall once again begin by discussing a campaign intended to force a major decision.

1. The immediate object of an attack is victory. Only by means of his superior strength can the attacker make up for all the advantages

that accrue to the defender by virtue of his position, and possibly by the modest advantage that his army derives from the knowledge that it is on the attacking, the advancing side. Usually this latter is much overrated: it is short-lived and will not stand the test of serious trouble. Naturally we assume that the defender will act as sensibly and correctly as the attacker. We say this in order to exclude certain vague notions about sudden assaults and surprise attacks, which are commonly thought of as bountiful sources of victory. They will only be that under exceptional circumstances. We have already discussed elsewhere the nature of a genuine strategic surprise.

If an attack lacks material superiority, it must have moral superiority to make up for its inherent weakness. Where even moral superiority is lacking, there is no point in attacking at all, for one cannot expect to succeed.

2. Prudence is the true spirit of defence, courage and confidence the true spirit of attack. Not that either form can do without both qualities, but each has a stronger affinity with one of them. After all, these qualities are necessary only because action is no mathematical construction, but has to operate in the dark, or at best in twilight. Trust must be placed in the guide whose qualifications are best suited to our purposes. The lower the defender's morale, the more daring the attacker should be.

3. Victory presupposes a clash of the two main forces. This presents less uncertainty to the attacker. His role is to confront the defender, whose positions are usually already known. In our discussion of the defence, on the other hand, we argued that if the defender has chosen a poor position the attacker should not seek him out, because the defender would have in that case to seek *him* out instead, and he would then have the advantage of catching the defender unprepared. In that case, everything would depend on the most important road and its general direction. This point was not discussed in the previous book, but was left until this chapter. We must therefore examine it now.

4. The possible objectives of an attack, and, consequently, the *aims* of victory, have already been discussed. If these lie within the theatre of war that we intend to attack, and within the probable sphere of victory, the natural direction of the blow will be determined by the roads leading to them. But one should not forget that the object of the attack usually gains significance only with victory;

victory must always be conceived in conjunction with it. So the attacker is not interested simply in reaching the objective: he must get there as victor. Consequently, his blow must be aimed not just at the objective but at the road that the enemy will have to take to reach it. The road then becomes the first objective. Victory can be made more complete if we encounter the enemy before he has reached that objective, cutting him off from it and getting there first. If for instance the main objective of the attack is the enemy's capital and the defender has not taken up a position between it and the attacker, the latter would be making a mistake if he advanced straight on the city. He would do better to strike at the communications between the enemy army and its capital and there seek the victory which will bring him to the city.

If there is no major objective within the area affected by the victory, the point of paramount importance is the enemy's line of communication with the nearest significant objective. Every attacker, therefore, has to ask himself how he will exploit his victory after the battle. The next objective to be won will then indicate the natural direction of his blow. If the defender has taken up his new position in that area he has made the correct choice, and the attacker has got to seek him out there. If that position is too strong, the attacker must try to by-pass it, making a virtue of necessity. But if the defender is not where he ought to be, the attacker must move in that direction himself. As soon as he is level with the defender— assuming the latter has made no lateral movement in the mean-time—he should wheel toward the enemy's lines of communication with the proper objective of seeking out his enemy there. If the latter has not moved at all, the attacker will have to turn and take him in the rear.

Among the roads from which the attacker may choose, the great commercial highways are the most obvious and suitable. But wherever they form too large a detour, one should take a more direct, even if a narrower road. A line of retreat that deviates considerably from a straight line always involves a serious risk.

5. An attacker bent on a major decision has no reason whatever to divide his forces. If in fact he does so, it may usually be ascribed to a state of confusion. His columns should advance on no wider a front than will allow them to be brought into action simultaneously. If the enemy force is divided, so much the better; in that case,

minor diversions are in order—strategic feints, made with the object of maintaining one's advantage. Should the attacker choose to divide his forces for that purpose he would be quite justified in doing so.

The division of the army into several columns, which in any case is indispensable, must be the basis for envelopment in the tactical attack; for envelopment is the most natural form of attack, and should not be disregarded without good cause. But the envelopment must be tactical; a strategic envelopment concurrent with a major blow is a complete waste of strength. It can only be justified if the attacker is strong enough not to have any doubts about the outcome.

6. But attack also requires caution: the attacker himself has a rear and communications to protect. This protection should, if possible, consist in the direction of advance—that is, it should be automatically provided by the army itself. If forces have to be detached for this purpose, thus causing a diversion of strength, it can only lessen the impact of the blow. A large army always advances on a front at least a day's march in width; so if the lines of communication and retreat do not deviate too much from the perpendicular, the front itself usually provides all the cover necessary.

Dangers of this sort to which the attacker is exposed can be gauged chiefly by the enemy's character and situation. If everything is subordinated to the pressure of an imminent major decision, the defender will have little scope for auxiliary operations, and the attacker, therefore, will not ordinarily be in great danger. But once the advance is over and the attacker gradually goes over to a state of defence, the protection of the rear assumes increasing urgency and importance. The attacker's rear is inherently more vulnerable than the defender's; so the latter may have started operations against the attacker's lines of communication long before he goes over to an actual offensive, and even while he is still on the retreat.

CHAPTER 16

ATTACK ON A THEATRE OF WAR: NOT SEEKING A DECISION

1. Even where determination and strength will not suffice to bring about a great decision, one may still want to mount a strategic attack against a minor objective. If that attack succeeds and the objective is attained, the situation reverts to a state of rest and balance. If difficulties are encountered to any serious extent, the advance is halted at an earlier stage. It will then be replaced either by offensives of opportunity or by mere strategic manoeuvre. That is the nature of most campaigns.

2. Objectives of such an offensive may be:

a. *A stretch of territory.* This may yield food-supplies; possibly contributions; protection of one's own territory; or a bargaining counter in peace negotiations. Sometimes the concept of military glory may play a part, as it constantly did in the campaigns fought by the French marshals under Louis XIV. The essential distinction lies in whether the territory can be held. As a rule, that is possible only if it borders on one's own theatre of operations and forms a natural extension to it. Only this type can constitute a bargaining counter at the peace table; all others are usually held temporarily, for the duration of the campaign, to be abandoned in the winter.

b. *An important depot.* If it were not important, it would hardly be considered an objective for an offensive taking up a whole campaign. It may in itself constitute a loss to the defender and a gain to the attacker; but the chief advantage to the latter lies in the fact that it will force the defender to withdraw and abandon territory which he would otherwise have held. Thus the capture of the depot is actually more of a means, and is listed here as an end only because it is the nearest immediate objective of action.

c. *The capture of a fortress.* We refer the reader to the separate chapter devoted to the capture of fortresses. It is clear from the arguments developed there why fortresses have always been the preferred and most desirable objectives in offensives or campaigns that could aim *neither* at the enemy's total defeat *nor* at the seizure of an important part of his country. So it is easily explained why in a country like the Netherlands, which is full of fortresses, the aim of operations has always been the capture of one fortress or another, the eventual seizure of the whole area *rarely emerging as the objective of the campaign*. Each fortress was deemed a discrete unit, and prized for its own sake. Apparently more attention was paid to the convenience and ease of the enterprise than to the actual value of the place.

Still, the siege of a fortress of any size is always an important operation because it is very expensive—an important consideration in wars that are not fought for major issues. That is why such a siege must be included among the significant elements of a strategic attack. The less important the place, the less determined the siege, the fewer the preparations made for it, the greater the likelihood of an air of improvisation, then the more the strategic objective will shrink in significance, and the weaker the forces and intentions to which it is suited. Such cases often end up as shadow-boxing, simply aimed at terminating the campaign honourably: as the attacker, one is after all bound to do something.

d. *A successful engagement, encounter*, or even *battle*, whether for the sake of trophies, or possibly simply of honour, and at times merely to satisfy a general's ambition. Anyone who doubts that this occurs does not know military history. Most of the offensive battles in the French campaigns during the age of Louis XIV* were of this type. It is more important to note, however, that these considerations are not without weight, mere quirks of vanity: they have a very definite bearing on the peace and hence they lead fairly straight to the goal. Military honour and the renown of an army and its generals are factors that operate invisibly, but they constantly permeate all military activity.

Such engagements, to be sure, are based on the following assumptions that: (a) there is a fair prospect of victory; and (b) if they end in defeat, not too much is lost. One must be careful not to confuse this type of battle, fought under restricted conditions for limited objectives, with victories that were not followed up for want of moral fibre.

3. With the exception of the last of these categories, (d), all can be achieved without major engagements. The means that the offensive can use for this purpose derive from the interests that the defender has to protect in his theatre of war. They will, therefore, consist in threatening his lines of communication, with its depots, rich provinces, important towns, or key points such as bridges, passes, etc.; or in the occupation of strong positions uncomfortably located for the defender; or in the occupation of important towns, fertile agricultural areas, or disaffected districts which can be seduced into revolt; or in threatening his weaker allies, and so on. If the attacker manages to disrupt communications to the point where the enemy cannot restore them without serious loss, if he sets out to seize these points, he will force the defender to take up another position to the rear or to the flank so as to cover them, even if it means giving up lesser ones. Thus an area is left uncovered, or a depot or fortress exposed—the former open to conquest, the latter to siege. Major or minor engagements may result, but they will neither be sought, nor will they be treated as objectives in themselves, but rather as necessary evils. They cannot rise above a certain level of magnitude and importance.

4. An operation on the part of a defender against the attacker's lines of communication is a type of reaction which, in a war aiming at major decisions, can take place only if those lines become very long. But in wars not seeking great decisions this type of reaction is more appropriate. Admittedly the enemy's lines will rarely be very extended, but the point here is not to inflict severe damage on him. It will often be enough to harass him and keep him short of supplies; and what the lines lack in length is to some extent made up by the length of time that can be spent on this kind of fighting. That is why the cover of his strategic flanks is of great importance to the attacker. If this kind of contest or rivalry develops between the attacker and the defender, the former will have to make up for his natural disadvantages by means of his superior numbers. If his strength and determination are still enough for him to risk a decisive blow at an enemy unit or even at the main enemy force, this threat, held over the defender's head, remains his best way of covering himself.

5. In conclusion, we must mention one other important advantage which the attacker enjoys in this sort of war: he is better placed to gauge the enemy's intentions and resources than the defender is to

gauge his. It is a great deal harder to predict the degree of vigour and daring with which the attacker will act than it is to predict whether the defender is contemplating a major stroke. In practice, the mere choice of the defensive form of warfare generally assures a lack of positive intentions. Besides, the difference between preparations for a major counterstroke and ordinary means of defence is much more marked than that between the preparations for a major attack and for a minor one. Finally, the defender is forced to make his dispositions earlier, thus giving the attacker the advantage of a counter-riposte.

CHAPTER 21

INVASION

ALMOST all we wish to say about invasion consists in a definition of the term. It is often used by modern writers—indeed, even with the air of designating a special quality. The French are always writing about *guerre d'invasion*. What they understand by it is any attack that penetrates deep into enemy territory, and they would like if possible to establish its meaning as the opposite of a routine attack—that is, one that merely nibbles at a frontier. That, however, is unscientific linguistic confusion. Whether an attack will halt at the frontier or penetrate into the heart of the enemy's territory, whether its main concern is to seize the enemy's fortresses or to seek out the core of enemy resistance and pursue it relentlessly, is not a matter that depends on form: it depends on circumstances. Theory, at least, permits no other answer. In some cases it may be more methodical and even more prudent to penetrate some distance rather than stay close to the frontier, but usually this is nothing but the successful outcome of a vigorous *attack*, and so cannot be distinguished from it in any way.

CHAPTER 22

THE CULMINATING POINT OF VICTORY*

IT is not possible in every war for the victor to overthrow his enemy completely. Often even victory has a culminating point. This has been amply demonstrated by experience. Because the matter is particularly important in military theory and forms the keystone for most plans of campaign, and because its surface is distorted by apparent contradictions, like the dazzling effect of brilliant colours, we shall examine it more closely and seek out its inner logic.

Victory normally results from the superiority of one side; from a greater aggregate of physical and psychological strength. This superiority is certainly augmented by the victory, otherwise it would not be so coveted or command so high a price. That is an automatic consequence of victory *itself*. Its effects exert a similar influence, but only up to a point. That point may be reached quickly—at times so quickly that the total consequences of a victorious battle may be limited to an increase in psychological superiority alone. We now propose to examine how that comes about.

As a war unfolds, armies are constantly faced with some factors that increase their strength and with others that reduce it. The question therefore is one of superiority. Every reduction in strength on one side can be considered as an increase on the other. It follows that this two-way process is to be found in attack as well as in defence.

What we have to do is examine the principal cause of this change in one of these instances, and so at the same time determine the other.

In an advance, the principal causes of additional strength are:

1. The losses suffered by the defending forces are usually heavier than those of the attacker.
2. The defender's loss of fixed assets such as magazines, depots, bridges, and the like, is not experienced by the attacker.
3. The defender's loss of ground, and therefore of resources, from the time we enter his territory.
4. The attacker benefits from the use of some of these resources; in other words, he can live at the enemy's expense.

5. The enemy loses his inner cohesion and the smooth functioning of all components of his force.

6. Some allies are lost to the defender, others turn to the invader.

7. Finally, the defender is discouraged, and so to some extent disarmed.

The causes of loss in strength for an invading army are:

1. The invader has to besiege, assault or observe the enemy's fortresses; while the defender, if he has previously been doing the same, will now add the units so employed to his main force.

2. The moment an invader enters enemy territory, the nature of the operational theatre changes. It becomes hostile. It must be garrisoned, for the invader can control it only to the extent that he has done so; but this creates difficulties for the entire machine, which will inevitably weaken its effectiveness.

3. The invader moves away from his sources of supply, while the defender moves closer to his own. This causes delay in the replacement of his forces.

4. The danger threatening the defender will bring allies to his aid.

5. Finally, the defender, being in real danger, makes the greater effort, whereas the efforts of the victor slacken off.

All these advantages and disadvantages may coexist; they can meet, so to speak, and pursue their ways in opposite directions. Only the last meet as true opposites: they cannot by-pass one another, so they are mutually exclusive. That alone is enough to show the infinite range of effects a victory can have—depending on whether they stun the loser or rouse him to greater efforts.

We shall try to qualify each of the above points in a few brief comments.

1. The enemy's losses may be at their maximum directly after his defeat, and then diminish daily until the point is reached where his strength equals ours. On the other hand, his losses may grow progressively day by day. All depends on differences in the overall situation and circumstances. Generally speaking, one can only say that the former is more likely to occur with a good army, the latter with a bad one. The most important factor besides the spirit of the troops is the spirit of the government. It is vital in war to distinguish between the two,

or one may stop at the very point where one should really start, and vice versa.

2. The enemy's loss of fixed assets may decrease or increase in the same way, depending on the location and nature of his supply depots. Nowadays, incidentally, this point is no longer so important as the others.

3. The third advantage cannot fail to grow with the progress of the advance. Indeed one can say that it only begins to count when the attack has penetrated deep into enemy territory— when a third or at least a quarter has been taken. A further factor is an area's intrinsic value in relation to the war effort.

4. The fourth advantage is also bound to increase as the advance proceeds.

In connection with these two last points, it should be noted that they seldom have an immediate effect on troops in action. Their work is slow and indirect. Therefore one should not on their account make too great an effort and so place oneself in too dangerous a situation.

5. The fifth advantage also only begins to tell after an army has advanced some distance, and when the configuration of the enemy's country provides an opportunity to isolate certain areas from the rest. Like tightly constricted branches, these will then tend to wither away.

6. and 7. It is probable, at all events, that the sixth and seventh advantages will increase with the advance. We shall return to them later on.

Now let us turn to the causes for loss in strength.

1. In most cases as an advance proceeds, there will be more sieges, assaults, and investments of fortresses. This on its own is so debilitating to *the available fighting forces* that it may easily cancel out all other advantages. True, in modern times one has begun to assault fortresses with very few troops and to observe them with still smaller numbers, and the enemy has, of course, to find garrisons for them. Nevertheless, fortresses remain an important element of security. Half the garrisons may usually consist of men who have not so far taken part in the war; yet one must still leave twice their strength in front of the fortresses on one's line of communication; and if even a single

important place has to be formally besieged or starved out, it will call for a small army.

2. The second cause of weakening, the establishment of a theatre of operations in enemy territory, grows, of course, with the advance. It may not immediately deplete the strength of the forces, but in the long run it will be even more effective than the first factor.

The only parts of enemy territory one can treat as being within one's theatre of operations are those one has actually occupied—by leaving small units in the field, by intermittent garrisons stationed in the major towns, by units established at the relay stations, and so on. Small as each of these garrisons may be, they all deplete the army's fighting strength. But that is the least important part.

Every army has strategic flanks—that is, the areas along both sides of its lines of communication; but since the enemy's army has the same, these are hardly considered a source of weakness. That, however, only holds true in one's own country. Once on enemy soil the weakness becomes palpable. If a long line of communications is covered poorly or not at all, the smallest operation against it holds out promise of success; and in enemy territory raiders may appear from any quarter.

The further the advance, the longer these flanks become, and the risks they represent will progressively increase. Not only are they hard to cover, but the very length of unprotected lines of communication tends to challenge the enemy's spirit of enterprise; and the consequences their loss can have in the event of a retreat are very grave indeed.

All this contributes to place a new burden on an advancing army with every step it takes; so unless it started with exceptional superiority, it will find its freedom of action dwindling and its offensive power progressively reduced. In the end, it will feel unsure of itself and nervous about its situation.

3. The third factor, the distance from the sources that must send continual replacements for this steadily weakening army, will increase proportionately with the advance. In this respect a conquering army is like the light of a lamp; as the oil that feeds it sinks and draws away from the focus, the light diminishes until at last it goes out altogether.

It is true that the wealth of the conquered areas may mitigate this problem, but it can never eliminate it altogether. There are always things that must be supplied from home—especially men. *In general*, deliveries from enemy resources are neither so prompt nor so reliable as those from one's own. In an emergency, help takes longer to arrive, while misunderstandings and mistakes of all kinds cannot be brought to light and rectified so promptly.

If a monarch does not command his troops in person, as has become customary in recent wars, if he is no longer easily available, a new and very serious handicap arises from the loss of time involved in the transmission of messages. Even the widest powers conferred on a commander will not suffice to meet every contingency that may arise in his sphere of action.

4. The change in political alignments. If these changes, resulting from his victories, are likely to be to the disadvantage of the victor, they will probably be so in direct proportion to his advance—which is also the case if they are to his advantage. All depends on the existing political affiliations, interests, traditions, lines of policy, and the personalities of princes, ministers, favourites, mistresses, and so forth. The only general comment one can make is that after the defeat of a major power with lesser allies, these will quickly desert their leader. In this respect, the victor will then gain strength with every blow. If, on the other hand, the defeated state is smaller, protectors will appear much sooner if its very existence is threatened. Others who may have helped to endanger it will detach themselves if they believe that the success is becoming too great.

5. The increased resistance aroused in the enemy. Sometimes, stunned and panic-stricken, the enemy may lay down his arms, at other times he may be seized by a fit of enthusiasm: there is a general rush to arms, and resistance is much stronger after the first defeat than it was before. The information from which one must guess at the probable reaction include the character of the people and the government, the nature of the country, and its political affiliations.

The last two points alone can make an infinite difference to the plans that one can and must make in war to take account of either possibility. While one man may lose his best chance through timidity and following so-called orthodox procedures, another will plunge in head first and end up looking as dazed and surprised as if he had just been fished out of the water.

Further, one should be conscious of the slackening of effort that not infrequently occurs on the part of the victor after the danger has been overcome, and when, on the contrary, fresh efforts are called for to follow up the victory. If we take an overall view of these differing and opposing principles, we will doubtless conclude that the utilization of the victory, a continued advance in an offensive campaign, will usually swallow up the superiority with which one began or which was gained by the victory.

At this point we are bound to ask: if all this is true, why does the winner persist in pursuing his victorious course, in advancing his offensive? Can one really still call this a 'utilization of victory'? Would he not do better to stop before he begins to lose the upper hand?

The obvious answer is that superior strength is not the end but only the means. The end is either to bring the enemy to his knees or at least to deprive him of some of his territory—the point in that case being *not to improve the current military position* but to improve one's general prospects in the war and in the peace negotiations. Even if one tries to destroy the enemy completely, one must accept the fact that every step gained may weaken one's superiority— though it does not necessarily follow that it must fall to zero before the enemy capitulates. He may do so at an earlier point, and if this can be accomplished with one's last ounce of superiority, it would be a mistake not to have used it.

Thus the superiority one has or gains in war is only the means and not the end; it must be risked for the sake of the end. But one must know the point to which it can be carried in order not to overshoot the target; otherwise instead of gaining new advantages, one will disgrace oneself.

There is no need to cite historical examples in order to prove that this is how loss of superiority affects a strategic attack. Indeed, such instances occur so frequently that we have felt it necessary to investigate their underlying causes. Only with the rise of Bonaparte have there been campaigns between civilized states where superiority has consistently led to the enemy's collapse. Before his time, every campaign had ended with the winning side attempting to reach a state of balance in which it could maintain itself. At that point, the progress of victory stopped, and a retreat might even be called for. This culminating point in victory is bound to recur in every future war in which the destruction of the enemy cannot be the military aim, and this will presumably be true of most wars. The natural goal of all campaign plans, therefore, is the turning point at which attack becomes defence.

If one were to go beyond that point, it would not merely be a *useless* effort which could not add to success. It would in fact be a *damaging* one, which would lead to a reaction; and experience goes to show that such reactions usually have completely disproportionate effects. This is such a universal experience, and appears so natural and easy to understand, that there is no need for a laborious investigation of its causes. The main causes are always the lack of organization in newly occupied territory, and the psychological effect of the stark contrast between the serious losses sustained and the successes that had been hoped for. There is an unusually active interplay between the extremes of morale—on the one hand, encouragement often verging on bravado, and on the other, depression. As a result, losses will be heavier during a retreat, and one can usually be grateful if one has to sacrifice only conquered territory, and not one's native soil.

At this point we must eliminate an apparent inconsistency.

This rests on the assumption that so long as an attack progresses there must still be some superiority on its side; further, that since defence (the more effective form of war) must start when the advance ends, one may not really be in much danger of imperceptibly becoming the weaker side. Yet that is what happens; history forces us to admit that the risk of a setback often does not reach its peak until the moment when the attack has lost its impetus and is turning into defence. We must look for the reason.

The superiority that I have attributed to the defensive form of warfare rests on the following:

1. The utilization of terrain
2. The possession of an organized theatre of operations
3. The support of the population
4. The advantage of being on the waiting side.

It is obvious that these factors will not everywhere be found in equal strength, or always be equally effective. One defence is therefore not exactly like another, nor will defence always enjoy the same degree of superiority over attack. In particular this will be the case in a defence that follows directly the exhaustion of an offensive—a defence whose theatre of operations is located at the apex of an offensive wedge thrust forward deep into hostile territory. Only the first of the four factors listed above, the utilization of terrain, will remain unchanged in such a defence; the second is usually eliminated, the third works in reverse, and the fourth is much reduced in strength. A word or two in explanation of this last point may be useful.

In an imaginary equilibrium, whole campaigns might often end without result because the side that should take the initiative lacks determination. That, in our view, is exactly why it is an advantage to be able to await the enemy. But if an offensive act upsets this equilibrium, damages the enemy's interests and impels him into action, he is far less likely to remain inactive and irresolute. A defence is far more provocative in character when it is undertaken on occupied territory than it is on one's own; it is, so to speak, infected with the virus of attack, and this weakens its basic character. In Silesia and Saxony Daun granted Frederick a period of calm that he would never have allowed in Bohemia.*

It is clear, therefore, that a defence that is undertaken in the framework of an offensive is weakened in all its key elements. It will thus no longer possess the superiority which basically belongs to it.

Just as no defensive campaign consists simply of defensive elements, so no offensive campaign consists purely of offensive ones. Apart from the short intervals in every campaign during which both sides are on the defensive, every attack which does not lead to peace must necessarily end up as a defence.

It is thus defence itself that weakens the attack. Far from this being idle sophistry, we consider it to be the greatest disadvantage of the attack that one is eventually left in a most awkward defensive position.

This will explain why there is a gradual reduction in the difference between the original effectiveness of attack and defence as forms of warfare. We now propose to show how this difference can for a time vanish altogether and reverse itself completely.

We can be more succinct if we may use an analogy from nature. Every physical force requires time to become effective. A force that, if gently and gradually applied, would suffice to arrest a body in motion, will be overcome by it if there is not enough time for it to operate. This law of physics provides a pertinent image of many features of our own psychology. Once our train of thought is set in a certain direction, many reasons which would otherwise be basically adequate to do so will not be able to deflect or arrest it. Time, repose, and a sustained impact on one's consciousness are needed. It is the same in war. Once the mind is set on a certain course toward its goal, or once it has turned back toward a refuge, it may easily happen that arguments which would compel one man to stop, and justify another in acting, will not easily be fully appreciated. Meanwhile the action continues, and in the sweep of motion one crosses the threshold of equilibrium, the line of culmination, without knowing it. It is even possible that the attacker, reinforced by the psychological forces peculiar to attack, will in spite of his exhaustion find it less difficult to go on than to stop—like a horse pulling a load uphill. We believe that this demonstrates without inconsistency how an attacker can overshoot the point at which, if he stopped and assumed the defensive, there would still be a chance of success—that is, of equilibrium. It is therefore important to calculate this point correctly when planning the campaign. An attacker may otherwise take on more than he can manage and, as it were, get into debt; a defender must be able to recognize this error if the enemy commits it, and exploit it to the full.

In reviewing the whole array of factors a general must weigh before making his decision, we must remember that he can gauge the direction and value of the most important ones only by considering numerous other possibilities—some immediate, some remote. He must *guess*, so to speak: guess whether the first shock of battle will steel the enemy's resolve and stiffen his resistance, or whether, like a Bologna flask, it will shatter as soon as its surface is scratched; guess the extent of debilitation and paralysis that the drying up of particular sources of supply and the severing of certain lines of communication will cause in the enemy; guess whether the burning pain of the injury he has been dealt will make the enemy collapse with exhaustion or, like a wounded bull, arouse his rage; guess whether the other powers will be frightened or indignant, and whether and which political alliances will be dissolved or formed. When we realize that he must hit upon all this and much more by means of his discreet judgement, as a marksman hits a target, we must admit that such an accomplishment of the human mind is no small achievement. Thousands of wrong turns running in all directions tempt his perception; and if the range, confusion and complexity of the issues are not enough to overwhelm him, the dangers and responsibilities may.

This is why the great majority of generals will prefer to stop well short of their objective rather than risk approaching it too closely, and why those with high courage and an enterprising spirit will often overshoot it and so fail to attain their purpose. Only the man who can achieve great results with limited means has really hit the mark.

BOOK EIGHT
WAR PLANS

CHAPTER I

INTRODUCTION

IN the chapter on the nature and purpose of war we roughly sketched the general concept of war and alluded to the connections between war and other physical and social phenomena, in order to give our discussion a sound theoretical starting point. We indicated what a variety of intellectual obstacles besets the subject, while reserving detailed study of them until later; and we concluded that the grand objective of all military action is to overthrow the enemy—which means destroying his armed forces. It was therefore possible to show in the following chapter that battle is the one and only means that warfare can employ. With that, we hoped, a sound working hypothesis had been established.

Then we examined, one by one, the salient patterns and situations (apart from battle itself) that occur in warfare, trying to gauge the value of each with greater precision, both according to its inherent characteristics and in the light of military experience. We also sought to strip away the vague, ambiguous notions commonly attached to them, and tried to make it absolutely clear that the destruction of the enemy is what always matters most.

We now revert to warfare as a whole, to the discussion of the planning of a war and of a campaign, which means returning to the ideas put forward in Book One.

The chapters that follow will deal with the problem of war as a whole. They cover its dominant, its most important aspect: pure strategy. We enter this crucial area—the central point on which all other threads converge—not without some diffidence. Indeed, this diffidence is amply justified.

On the one hand, military operations appear extremely simple. The greatest generals discuss them in the plainest and most forthright language; and to hear them tell how they control and manage that enormous, complex apparatus one would think the only thing that mattered was the speaker, and that the whole monstrosity called war came down, in fact, to a contest between individuals, a sort of duel. A few uncomplicated thoughts seem to account for their decisions—either that, or the explanation lies in various emotional states; and one is left with the impression that great commanders manage matters in an easy, confident and, one would almost think, off-hand sort of way. At the same time we can see how many factors are involved and have to be weighed against each other; the vast, the almost infinite distance there can be between a cause and its effect, and the countless ways in which these elements can be combined. The function of theory is to put all this in systematic order, clearly and comprehensively, and to trace each action to an adequate, compelling cause. When we contemplate all this, we are overcome by the fear that we shall be irresistibly dragged down to a state of dreary pedantry, and grub around in the underworld of ponderous concepts where no great commander, with his effortless *coup d'oeil*, was ever seen. If that were the best that theoretical studies could produce it would be better never to have attempted them in the first place. Men of genuine talent would despise them and they would quickly be forgotten. When all is said and done, it really is the commander's *coup d'oeil*, his ability to see things simply, to identify the whole business of war completely with himself, that is the essence of good generalship. Only if the mind works in this comprehensive fashion can it achieve the freedom it needs to dominate events and not be dominated by them.

We resume our task then, with some diffidence; and we shall fail unless we keep to the path we set ourselves at the beginning. Theory should cast a steady light on all phenomena so that we can more easily recognize and eliminate the weeds that always spring from ignorance; it should show how one thing is related to another, and keep the important and the unimportant separate. If concepts combine of their own accord to form that nucleus of truth we call a principle, if they spontaneously compose a pattern that becomes a rule, it is the task of the theorist to make this clear.

The insights gained and garnered by the mind in its wanderings among basic concepts are benefits that theory can provide. Theory cannot equip the mind with formulas for solving problems, nor can it mark the narrow path on which the sole solution is supposed to lie by planting a hedge of principles on either side. But it can give the mind insight into the great mass of phenomena and of their relationships, then leave it free to rise into the higher realms of action. There the mind can use its innate talents to capacity, combining them all so as to seize on what is *right* and *true* as though this were a single idea formed by their concentrated pressure—as though it were a response to the immediate challenge rather than a product of thought.

CHAPTER 2

ABSOLUTE WAR AND REAL WAR*

WAR plans cover every aspect of a war, and weave them all into a single operation that must have a single, ultimate objective in which all particular aims are reconciled. No one starts a war—or rather, no one in his senses ought to do so—without first being clear in his mind what he intends to achieve by that war and how he intends to conduct it. The former is its political purpose; the latter its operational objective. This is the governing principle which will set its course, prescribe the scale of means and effort which is required, and make its influence felt throughout down to the smallest operational detail.

We said in the opening chapter that the natural aim of military operations is the enemy's overthrow, and that strict adherence to the logic of the concept can, in the last analysis, admit of no other. Since both belligerents must hold that view it would follow that military operations could not be suspended, that hostilities could not end until one or other side were finally defeated.

In the chapter on the suspension of military activity we showed how factors inherent in the war-machine itself can interrupt and modify the principle of enmity as embodied in its agent, man, and in all that goes to make up warfare. Still, that process of modification is by no means adequate to span the gap between the pure concept of war and the concrete form that, as a general rule, war assumes. Most wars are like a flaring-up of mutual rage, when each party takes up arms in order to defend itself, to overawe its opponent, and occasionally to deal him an actual blow. Generally it is not a case in which two mutually destructive elements collide, but one of tension between two elements, separate for the time being, which discharge energy in discontinuous, minor shocks.

But what exactly is this nonconducting medium, this barrier that prevents a full discharge? Why is it that the theoretical concept is not fulfilled in practice? The barrier in question is the vast array of factors, forces and conditions in national affairs that are affected by war. No logical sequence could progress through their innumerable twists and turns as though it were a simple thread that linked two deductions. Logic comes to a stop in this labyrinth; and those men who habitually act, both in great and minor affairs, on particular dominating impressions or feelings rather than according to strict logic, are hardly aware of the confused, inconsistent, and ambiguous situation in which they find themselves.

The man in overall command may actually have examined all these matters without losing sight of his objective for an instant; but the many others concerned cannot all have achieved the same insight. Opposition results, and in consequence something is required to overcome the vast inertia of the mass. But there is not usually enough energy available for this.

This inconsistency can appear in either belligerent party or in both, and it is the reason why war turns into something quite different from what it should be according to theory—turns into something incoherent and incomplete.

This is its usual appearance, and one might wonder whether there is any truth at all in our concept of the absolute character of war were it not for the fact that with our own eyes we have seen warfare achieve this state of absolute perfection. After the short prelude of the French Revolution, Bonaparte brought it swiftly and ruthlessly to that point. War, in his hands, was waged without respite until the enemy succumbed, and the counter-blows were struck with almost equal energy. Surely it is both natural and inescapable that this phenomenon should cause us to turn again to the pure concept of war with all its rigorous implications.

Are we then to take this as the standard, and judge all wars by it, however much they may diverge? Should we deduce our entire theory from it? The question is whether that should be the only kind of war or whether there can be other valid forms. We must make up our minds before we can say anything intelligent about war plans.

If the first view is right, our theory will everywhere approximate to logical necessity, and will tend to be clear and unambiguous. But in that case, what are we to say about all the wars that have been fought since the days of Alexander*—excepting certain Roman campaigns—down to Bonaparte? We should have to condemn them outright, but might be appalled at our presumption if we did so.

Worse still, we should be bound to say that in spite of our theory there may even be other wars of this kind in the next ten years, and that our theory, though strictly logical, would not apply to reality. We must, therefore, be prepared to develop our concept of war as it ought to be fought, not on the basis of its pure definition, but by leaving room for every sort of extraneous matter. We must allow for natural inertia, for all the friction of its parts, for all the inconsistency, imprecision, and timidity of man; and finally we must face the fact that war and its forms result from ideas, emotions, and conditions prevailing at the time—and to be quite honest we must admit that this was the case even when war assumed its absolute state under Bonaparte.

If this is the case, if we must admit that the origin and the form taken by a war are not the result of any ultimate resolution of the vast array of circumstances involved, but only of those features that happen to be dominant, it follows that war is dependent on the interplay of possibilities and probabilities, of good and bad luck, conditions in which strictly logical reasoning often plays no part at all and is always apt to be a most unsuitable and awkward intellectual tool. It follows, too, that war can be a matter of degree.

Theory must concede all this; but it has the duty to give priority to the absolute form of war and to make that form a general point of reference, so that he who wants to learn from theory becomes accustomed to keeping that point in view constantly, to measuring all his hopes and fears by it, and to approximating it *when he can* or *when he must*.

A principle that underlies our thoughts and actions will undoubtedly lend them a certain tone and character, though the immediate causes of our action may have different origins, just as the tone a painter gives to his canvas is determined by the colour of the underpainting.

If theory can effectively do this today, it is because of our recent wars. Without the cautionary examples of the destructive power of war unleashed, theory would preach to deaf ears. No one would have believed possible what has now been experienced by all.

Would Prussia in 1792 have dared to invade France with 70,000 men if she had had an inkling that the repercussions in case of failure would be strong enough to overthrow the old European balance of power? Would she, in 1806, have risked war with France with

100,000 men, if she had suspected that the first shot would set off a mine that was to blow her to the skies?

CHAPTER 3

A. INTERDEPENDENCE OF THE ELEMENTS OF WAR

SINCE war can be thought of in two different ways—its absolute form or one of the variant forms that it actually takes—two different concepts of success arise.

In the absolute form of war, where everything results from necessary causes and one action rapidly affects another, there is, if we may use the phrase, no intervening neutral void. Since war contains a host of interactions[1] since the whole series of engagements is, strictly speaking, linked together,[2] since in every victory there is a culminating point beyond which lies the realm of losses and defeats[3]—in view of all these intrinsic characteristics of war, we say there is only one result that counts: *final victory*. Until then, nothing is decided, nothing won, and nothing lost. In this form of war we must always keep in mind that it is the end that crowns the work. Within the concept of absolute war, then, war is indivisible, and its component parts (the individual victories) are of value only in their relation to the whole. Conquering Moscow and half of Russia in 1812 was of no avail to Bonaparte unless it brought him the peace he had in view. But these successes were only a part of his plan of campaign: what was still missing was the destruction of the Russian army. If that achievement had been added to the rest, peace would have been as sure as things of that sort ever can be. But it was too late to achieve the second part of his plan; his chance had gone. Thus the successful stage was not only wasted but led to disaster.

Contrasting with this extreme view of the connection between successes in war, is another view, no less extreme; which holds that war consists of separate successes each unrelated to the next, as in a

---

[1] See Chapter One, Book One.
[2] See Chapter Two, Book One.
[3] See Chapters Four and Five, Book Seven.

match consisting of several games. The earlier games have no effect upon the later. All that counts is the total score, and each separate result makes its contribution toward this total.

The first of these two views of war derives its validity from the nature of the subject; the second, from its actual history. Countless cases have occurred where a small advantage could be gained without an onerous condition being attached to it. The more the element of violence is moderated, the commoner these cases will be; but just as absolute war has never in fact been achieved, so we will never find a war in which the second concept is so prevalent that the first can be disregarded altogether. If we postulate the first of the two concepts, it necessarily follows from the start that every war must be conceived of as a single whole, and that with his first move the general must already have a clear idea of the goal on which all lines are to converge.

If we postulate the second concept, we will find it legitimate to pursue minor advantages for their own sake and leave the future to itself.

Since both these concepts lead to results, theory cannot dispense with either. Theory makes this distinction in the application of the two concepts: all action must be based on the former, since it is the fundamental concept; the latter can be used only as a modification justified by circumstances.

In 1742, 1744, 1757, and 1758,* when Frederick, operating from Silesia and Saxony, thrust new spearheads into Austria, he was well aware that they could not lead to another permanent acquisition such as Silesia and Saxony. His aim was not to overthrow the Austrian Empire but a secondary one, namely to gain time and strength. And he could pursue this secondary aim without any fear of risking his own existence.[4]

---

[4] If Frederick had won the battle of Kolin* and in consequence had captured the main Austrian army in Prague with both its senior commanders, it would indeed have been such a shattering blow that he might well have thought of pressing on to Vienna, shaking the foundations of the monarchy and imposing peace. That would have been an unparalleled success for those days, as great as the triumphs of the Napoleonic wars, but still more wonderful and brilliant for the disparity in size between the Prussian David and the Austrian Goliath. Victory at Kolin would almost certainly have made this success possible. But that does not invalidate the assertion made above, which only concerned the original purpose of the King's offensive. To surround and capture the enemy's main army, on the other hand, was something wholly unprovided for and the King had never given it a thought—at least until the Austrians invited it by the inadequate position they took up at Prague.*

However, when Prussia in 1806, and Austria in 1805 and 1809, adopted a still more modest aim—to drive the French across the Rhine—it would have been foolish if they had not begun by carefully reviewing the whole chain of events that success or failure would be likely to bring in consequence of the initial step, and which would lead to peace. Such a review was indispensable, both in order to decide how far they could safely exploit their successes and also how and where any enemy successes could be arrested.

Careful study of history shows where the difference between these cases lies. In the eighteenth century, in the days of the Silesian campaigns, war was still an affair for governments alone, and the people's role was simply that of an instrument. At the onset of the nineteenth century, peoples themselves were in the scale on either side. The generals opposing Frederick the Great were acting on instructions—which implied that caution was one of their distinguishing characteristics. But now the opponent of the Austrians and Prussians was—to put it bluntly—the God of War himself.

Such a transformation of war might have led to new ways of thinking about it. In 1805, 1806, and 1809 men might have recognized that total ruin was a possibility—indeed it stared them in the face. It might have stimulated them to different efforts that were directed toward greater objectives than a couple of fortresses and a medium-sized province.

They did not, however, change their attitude sufficiently, although the degree of Austrian and Prussian rearmament shows that the storm clouds massing in the political world had been observed. They failed because the transformations of war had not yet been sufficiently revealed by history. In fact the very campaigns of 1805, 1806, and 1809, and those that followed are the ones that make it easier for us to grasp the concept of modern, absolute war in all its devastating power.

Theory, therefore, demands that at the outset of a war its character and scope should be determined on the basis of the political probabilities. The closer these political probabilities drive war toward the absolute, the more the belligerent states are involved and drawn in to its vortex, the clearer appear the connections between its separate actions, and the more imperative the need not to take the first step without considering the last.

B. SCALE OF THE MILITARY OBJECTIVE AND OF THE EFFORT TO BE MADE

THE degree of force that must be used against the enemy depends on the scale of political demands on either side. These demands, so far as they are known, would show what efforts each must make; but they seldom are fully known—which may be one reason why both sides do not exert themselves to the same degree.

Nor are the situation and conditions of the belligerents alike. This can be a second factor.

Just as disparate are the governments' strength of will, their character and abilities.

These three considerations introduce uncertainties that make it difficult to gauge the amount of resistance to be faced and, in consequence, the means required and the objectives to be set.

Since in war too small an effort can result not just in failure but in positive harm, each side is driven to outdo the other, which sets up an interaction.

Such an interaction could lead to a maximum effort if a maximum could be defined. But in that case all proportion between action and political demands would be lost: means would cease to be commensurate with ends, and in most cases a policy of maximum exertion would fail because of the domestic problems it would raise.

In this way the belligerent is again driven to adopt a middle course. He would act on the principle of using no greater force, and setting himself no greater military aim, than would be sufficient for the achievement of his political purpose. To turn this principle into practice he must renounce the need for absolute success in each given case, and he must dismiss remoter possibilities from his calculations.

At this point, then, intellectual activity leaves the field of the exact sciences of logic and mathematics. It then becomes an art in the broadest meaning of the term—the faculty of using judgement to detect the most important and decisive elements in the vast array of facts and situations. Undoubtedly this power of judgement consists to a greater or lesser degree in the intuitive comparison of all the factors and attendant circumstances; what is remote and secondary is at once dismissed while the most pressing and important points are identified with greater speed than could be done by strictly logical deduction.

To discover how much of our resources must be mobilized for war, we must first examine our own political aim and that of the enemy. We must gauge the strength and situation of the opposing state. We must gauge the character and abilities of its government and people and do the same in regard to our own. Finally, we must evaluate the political sympathies of other states and the effect the war may have on them. To assess these things in all their ramifications and diversity is plainly a colossal task. Rapid and correct appraisal of them clearly calls for the intuition of a genius; to master all this complex mass by sheer methodical examination is obviously impossible. Bonaparte was quite right when he said that Newton* himself would quail before the algebraic problems it could pose.

The size and variety of factors to be weighed, and the uncertainty about the proper scale to use, are bound to make it far more difficult to reach the right conclusion. We should also bear in mind that the vast, unique importance of war, while not increasing the complexity and difficulty of the problem, does increase the value of the correct solution. Responsibility and danger do not tend to free or stimulate the average person's mind—rather the contrary; but wherever they do liberate an individual's judgement and confidence we can be sure that we are in the presence of exceptional ability.

At the outset, then, we must admit that an imminent war, its possible aims, and the resources it will require, are matters that can only be assessed when every circumstance has been examined in the context of the whole, which of course includes the most ephemeral factors as well. We must also recognize that the conclusion reached can be no more wholly objective than any other in war, but will be shaped by the qualities of mind and character of the men making the decision—of the rulers, statesmen, and commanders, whether these roles are united in a single individual or not.

A more general and theoretical treatment of the subject may become feasible if we consider the nature of states and societies as they are determined by their times and prevailing conditions. Let us take a brief look at history.

The semibarbarous Tartars, the republics of antiquity, the feudal lords and trading cities of the Middle Ages, eighteenth-century kings and the rulers and peoples of the nineteenth century—all conducted war in their own particular way, using different methods and pursuing different aims.

The Tartar hordes searched for new land. Setting forth as a nation, with women and children, they outnumbered any other army. Their aim was to subdue their enemies or expel them. If a high degree of civilization could have been combined with such methods, they would have carried all before them.

The republics of antiquity, Rome excepted, were small and their armies smaller still, for the *plebs*, the mass of the people, was excluded. Being so many and so close together these republics found that the balance that some law of nature will always establish among small and unconnected units formed an obstacle to major enterprises. They therefore limited their wars to plundering the countryside and seizing a few towns, in order to gain a degree of influence over them.

Rome was the one exception to this rule, and then only in her later days. With little bands of men, she had for centuries carried on the usual struggle with her neighbours for booty or alliance. She grew not so much by conquest as by the alliances she made, for the neighbouring peoples gradually merged with her and were assimilated into a greater Rome. Only when this process had spread the rule of Rome through Southern Italy did she begin to expand by way of actual conquest. Carthage fell; Spain and Gaul were taken; Greece was subjugated; and Roman rule was carried into Asia and Egypt. Rome's military strength at that period was immense, without her efforts being equally great. Her armies were kept up by her wealth. Rome was no longer like the Greek republics, nor was she even faithful to her own past. Her case is unique.

As singular in their own way were the wars of Alexander. With his small but excellently trained and organized army, Alexander shattered the brittle states of Asia. Ruthlessly, without pause, he advanced through the vast expanse of Asia until he reached India. That was something no republic could have achieved; only a king who in a sense was his own *condottiere** could have accomplished it so rapidly.

Mediaeval monarchs, great and small, waged war with feudal levies, which limited operations. If a thing could not be finished quickly it was impossible. The feudal army itself was an assemblage of vassals and their servants, brought and held together in part by legal obligation, in part by voluntary alliance—the whole amounting to true confederation. Weapons and tactics were based on individual combat, and thus unsuited to the organized action of large numbers. And indeed, cohesion in the state was never weaker or the individual so independent. It was the combination of these factors that gave mediaeval wars their special character. They were waged relatively quickly; not much time was wasted in the field; but their aim was usually to punish the enemy, not subdue him. When his cattle had been driven off and his castles burned, one could go home.

The great commercial cities and the small republics created *condottieri*. They were an expensive and therefore small military force. Even smaller was their fighting value: extremes of energy or exertion were conspicuous by their absence and fighting was generally a sham. In brief, then, hatred and enmity no longer drove the state to take matters into its own hands; they became an element in negotiation. War lost many of its risks; its character was wholly changed, and no deduction from its proper nature was still applicable.

Gradually the feudal system hardened into clearly delimited territorial sovereignty. States were closer knit; personal service was commuted into dues in kind, mostly in the form of money, and feudal levies were replaced by mercenaries. The transition was bridged by the *condottieri*. For a period they were also the instrument of the larger states. But soon the soldier hired on short-term contract evolved into the *permanent mercenary*, and the armed force of the state had become a standing army, paid for by the treasury.

The slow evolution toward this goal naturally brought with it numerous overlappings of these three types of military institutions. Under Henry IV* of France feudal levies, *condottieri* and a standing army were used side by side. The *condottieri* survived into the Thirty Years War, and indeed faint traces of them can be found in the eighteenth century.

Just as the character of the military institutions of the European states differed in the various periods, so did all their other conditions. Europe, essentially, had broken down into a mass of minor states. Some were turbulent republics, other precarious small monarchies with very limited central power. A state of that type could not be said to be genuinely united; it was rather an agglomeration of loosely associated forces. Therefore we should not think of such a state as a personified intelligence acting according to simple and logical rules.

This is the point of view from which the policies and wars of the Middle Ages should be considered. One need only think of the German emperors with their constant descents into Italy over a period of five hundred years. These expeditions never resulted in any complete conquest of the country; nor were they ever meant to do so. It would be easy to regard them as a chronic error, a delusion born of the spirit of the times; but there would be more sense in attributing them to a host of major causes, which we may possibly assimilate intellectually, but whose dynamic we will never comprehend as clearly as did the men who were actually obliged to contend with them. So long as the great powers that eventually grew out of this chaos needed time to consolidate and organize themselves, most of their strength and energies went into that process. Foreign wars were fewer, and those that did take place betrayed the marks of immature political cohesion.

The wars of the English against the French are the first to stand out. But France could not yet be considered as a genuine monarchy—she was rather an agglomeration of duchies and counties; while England, though displaying greater unity, still fought with feudal levies amid much domestic strife.

Under Louis XI* France took the greatest step toward internal unity. She became a conquering power in Italy under Charles VIII,* and her state and her army reached a peak under Louis XIV.*

Spanish unity began to form under Ferdinand of Aragon.* Under Charles V,* as a result of favourable marriages, a mighty Spanish monarchy suddenly emerged, composed of Spain and Burgundy, Germany and Italy. What this colossus lacked in cohesion and domestic stability was made up for by its wealth. Its standing army first encountered that of France. On the abdication of Charles V the colossus broke into two parts—Spain and Austria. The latter, strengthened by Hungary and Bohemia,* now emerged as a major power, dragging behind her the German confederation* like a dinghy.

The end of the seventeenth century, the age of Louis XIV, may be regarded as that point in history when the standing army in the shape familiar to the eighteenth century reached maturity. This military organization was based on money and recruitment. The states of Europe had achieved complete internal unity. With their subjects' services converted into money payments, the strength of governments now lay entirely in their treasuries. Thanks to cultural developments and to a progressively more sophisticated administration, their power was very great compared with earlier days. France put several hundred thousand regular troops in the field, and other states could do likewise in proportion to their populations.

International relations had changed in other ways as well. Europe was now split between a dozen monarchies and a handful of republics. It was conceivable that two states could fight a major war without, as in former times, involving twenty others. The possible political alignments were still many and various; but they could be surveyed, and their probability at each given instant could be evaluated.

Domestically almost every state had been reduced to an absolute monarchy; the privileges and influence of the estates had gradually disappeared. The executive had become completely unified and represented the state in its foreign relations. Political and military institutions had developed into an effective instrument, with which an independent will at the centre could now wage war in a form that matched its theoretical concept.

During this period, moreover, three new Alexanders appeared— Gustavus Adolphus,* Charles XII,* and Frederick the Great. With relatively limited but highly efficient forces each sought to turn his small state into a large monarchy, and crush all opposition. Had they been dealing only with Asiatic empires they might have resembled Alexander more closely. But in terms of risks that they ran, they undeniably foreshadowed Bonaparte.

But, if war gained in power and effectiveness, it lost in other respects.

Armies were paid for from the treasury, which rulers treated almost as their privy purse or at least as the property of the government, not of the people. Apart from a few commercial matters, relations with other states did not concern the people but only the treasury or the government. That at least was the general attitude. A government behaved as though it owned and managed a great estate that it constantly endeavoured to enlarge—an effort in which the inhabitants were not expected to show any particular interest. The Tartar people and army had been one; in the republics of antiquity and during the Middle Ages the people (if we confine the concept to those who had the rights of citizens) had still played a prominent part; but in the circumstances of the eighteenth century the people's part had been extinguished. The only influence the people continued to exert on war was an indirect one—through its general virtues or shortcomings.

War thus became solely the concern of the government to the extent that governments parted company with their peoples and behaved as if they were themselves the state. Their means of waging war came to consist of the money in their coffers and of such idle vagabonds as they could lay their hands on either at home or abroad. In consequence the means they had available were fairly well defined, and each could gauge the other side's potential in terms both of numbers and of time. War was thus deprived of its most dangerous feature—its tendency toward the extreme, and of the whole chain of unknown possibilities which would follow.

The enemy's cash resources, his treasury and his credit, were all approximately known; so was the size of his fighting forces. No great expansion was feasible at the outbreak of war. Knowing the limits of the enemy's strength, men knew they were reasonably safe from total ruin; and being aware of their own limitations, they were compelled to restrict their own aims in turn. Safe from the threat of extremes, it was no longer necessary to go to extremes. Necessity was no longer an incentive to do so, and the only impulse could come from courage and ambition. These, on the other hand, were strongly curbed by the prevailing conditions of the state. Even a royal commander had to use his army with a minimum of risk. If the army was pulverized, he could not raise another, and behind the army there was nothing. That enjoined the greatest prudence in all operations. Only if a decisive advantage seemed possible could the precious instrument be used, and to bring things to that point was a feat of the highest generalship. But so long as that was not achieved, operations drifted in a kind of vacuum; there was no reason to act, and every motivating force seemed inert. The original motive of the aggressor faded away in prudence and hesitation.

The conduct of war thus became a true game, in which the cards were dealt by time and by accident. In its effect it was a somewhat stronger form of diplomacy, a more forceful method of negotiation, in which battles and sieges were the principal notes exchanged. Even the most ambitious ruler had no greater aims than to gain a number of advantages that could be exploited at the peace conference.

This limited, constricted form of war was due, as we said, to the narrow base on which it rested. But the explanation why even gifted commanders and monarchs such as Gustavus Adolphus, Charles XII, and Frederick the Great, with armies of exceptional quality, should have risen so little above the common level of the times, why even they had to be content with moderate success, lies with the balance of power in Europe. With the multitude of minor states in earlier times, any one of them was prevented from rapidly expanding by such immediate and concrete factors as their proximity and contiguity, their family ties and personal acquaintances. But now that states were larger and their centres farther apart, the wide spread of interests they had developed became the factor limiting their growth. Political relations, with their affinities and antipathies, had become so sensitive a nexus that no cannon could be fired in Europe without every government feeling its interest affected. Hence a new Alexander needed more than his own sharp sword: he required a ready pen as well. Even so, his conquests rarely amounted to very much.

Even Louis XIV, though bent on destroying the balance of power in Europe and little troubled by the general hostility he faced by the end of the seventeenth century, continued waging war along traditional lines. While his military instrument was that of the greatest and richest monarch of all, its character was no different from that of his opponents'.

It had ceased to be in harmony with the spirit of the times to plunder and lay waste the enemy's land, which had played such an important role in antiquity, in Tartar days and indeed in mediaeval times. It was rightly held to be unnecessarily barbarous, an invitation to reprisals, and a practice that hurt the enemy's subjects rather than their government—one therefore that was ineffective and only served permanently to impede the advance of general civilization. Not only in its means, therefore, but also in its aims, war increasingly became limited to the fighting force itself. Armies, with their fortresses and prepared positions, came to form a state within a state, in which violence gradually faded away.

All Europe rejoiced at this development. It was seen as a logical outcome of enlightenment. This was a misconception. Enlightenment can never lead to inconsistency: as we have said before and shall have to say again, it can never make two and two equal five. Nevertheless this development benefited the peoples of Europe, although there is no denying that it turned war even more into the exclusive concern of governments and estranged it still further from the interests of the people. In those days, an aggressor's usual plan of war was to seize an enemy province or two. The defender's plan was simply to prevent him doing so. The plan for a given campaign was to take an enemy fortress or prevent the capture of one's own. No battle was ever sought, or fought, unless it were indispensable for that purpose. Anyone who fought a battle that was not strictly necessity, simply out of innate desire for victory, was considered reckless. A campaign was usually spent on a single siege, or two at the most. Winter quarters were assumed to be necessary for everyone. The poor condition of one side did not constitute an advantage to the other, and contact almost ceased between both. Winter quarters set strict limits to the operations of a campaign.

If forces were too closely balanced, or if the more enterprising side was also clearly the weaker of the two, no battle was fought and no town was besieged. The whole campaign turned on the retention of certain positions and depots and the systematic exploitation of certain areas.

So long as this was the general style of warfare, with its violence limited in such strict and obvious ways, no one saw any inconsistency in it. On the contrary, it all seemed absolutely right; and when in the eighteenth century critics began to analyse the art of war, they dealt with points of detail, without bothering much about fundamentals. Greatness, indeed perfection, was discerned in many guises, and even the Austrian Field-Marshal Daun—to whom it was mainly due that Frederick the Great completely attained his object and Maria Theresa* completely failed in hers—could be considered a great commander. Only from time to time someone of penetrating judgement—of real common sense—might suggest that with superior forces one should achieve positive results; otherwise the war, with all its artistry, was being mismanaged.

This was the state of affairs at the outbreak of the French Revolution. Austria and Prussia tried to meet this with the diplomatic type of war that we have described. They soon discovered its inadequacy. Looking at the situation in this conventional manner, people at first expected to have to deal only with a seriously weakened

French army; but in 1793 a force appeared that beggared all imagination. Suddenly war again became the business of the people—a
people of thirty millions, all of whom considered themselves to be
citizens. We need not study in detail the circumstances that accompanied this tremendous development; we need only note the effects
that are pertinent to our discussion. The people became a participant
in war; instead of governments and armies as heretofore, the full
weight of the nation was thrown into the balance. The resources and
efforts now available for use surpassed all conventional limits; nothing now impeded the vigour with which war could be waged, and
consequently the opponents of France faced the utmost peril.

The effects of this innovation did not become evident or fully
felt until the end of the revolutionary wars. The revolutionary quarrels did not yet advance inevitably toward the ultimate conclusion:
the destruction of the European monarchies. Here and there the
German armies were still able to resist them and stem the tide of
victory. But all this was really due only to technical imperfections
that hampered the French, and which became evident first in the
rank and file, then in their generals, and under the Directory in the
government itself.

Once these imperfections were corrected by Bonaparte, this jug-
gernaut of war, based on the strength of the entire people, began its
pulverizing course through Europe. It moved with such confidence
and certainty that whenever it was opposed by armies of the traditional type there could never be a moment's doubt as to the result.
Just in time, the reaction set in. The Spanish War spontaneously
became the concern of the people. In 1809 the Austrian government
made an unprecedented effort with reserves and militia; it came
within sight of success and far surpassed everything Austria had
earlier considered possible. In 1812 Russia took Spain and Austria
as models: her immense spaces permitted her measures—belated
though they were—to take effect, and even increased their effectiveness. The result was brilliant. In Germany, Prussia was first to rise.
She made the war a concern of the people, and with half her former
population, without money or credit, she mobilized a force twice as
large as she had in 1806. Little by little the rest of Germany followed
her example, and Austria too—though her effort did not equal that
of 1809—exerted an exceptional degree of energy. The result was
that in 1813 and 1814 Germany and Russia put about a million men

into the field against France—counting all who fought and fell in the two campaigns.

Under these conditions the war was waged with a very different degree of vigour. Although it did not always match the intensity of the French, and was at times even marked by timidity, campaigns were on the whole conducted in the new manner, not in that of the past. In the space of only eight months the theatre of operations changed from the Oder to the Seine. Proud Paris had for the first time to bow her head, and the terrible Bonaparte lay bound and chained.

Since Bonaparte, then, war, first among the French and subsequently among their enemies, again became the concern of the people as a whole, took on an entirely different character, or rather closely approached its true character, its absolute perfection. There seemed no end to the resources mobilized; all limits disappeared in the vigour and enthusiasm shown by governments and their subjects. Various factors powerfully increased that vigour: the vastness of available resources, the ample field of opportunity, and the depth of feeling generally aroused. The sole aim of war was to overthrow the opponent. Not until he was prostrate was it considered possible to pause and try to reconcile the opposing interests.

War, untrammelled by any conventional restraints, had broken loose in all its elemental fury. This was due to the peoples' new share in these great affairs of state; and their participation, in turn, resulted partly from the impact that the Revolution had on the internal conditions of every state and partly from the danger that France posed to everyone.

Will this always be the case in future? From now on will every war in Europe be waged with the full resources of the state, and therefore have to be fought only over major issues that affect the people? Or shall we again see a gradual separation taking place between government and people? Such questions are difficult to answer, and we are the last to dare to do so. But the reader will agree with us when we say that once barriers—which in a sense consist only in man's ignorance of what is possible—are torn down, they are not so easily set up again. At least when major interests are at stake, mutual hostility will express itself in the same manner as it has in our own day.

At this point our historical survey can end. Our purpose was not to assign, in passing, a handful of principles of warfare to each period. We wanted to show how every age had its own kind of war, its own limiting conditions, and its own peculiar preconceptions. Each period, therefore, would have held to its own theory of war, even if the urge had always and universally existed to work things out on scientific principles. It follows that the events of every age must be judged in the light of its own peculiarities. One cannot, therefore, understand and appreciate the commanders of the past until one has placed oneself in the situation of their times, not so much by a painstaking study of all its details as by an accurate appreciation of its major determining features.

But war, though conditioned by the particular characteristics of states and their armed forces, must contain some more general— indeed, a universal—element with which every theorist ought above all to be concerned.

The age in which this postulate, this universally valid element, was at its strongest was the most recent one, when war attained the absolute in violence. But it is no more likely that war will always be so monumental in character than that the ample scope it has come to enjoy will again be severely restricted. A theory, then, that dealt exclusively with absolute war would either have to ignore any case in which the nature of war had been deformed by outside influence, or else it would have to dismiss them all as misconstrued. That cannot be what theory is for. Its purpose is to demonstrate what war is in practice, not what its ideal nature ought to be. So the theorist must scrutinize all data with an inquiring, a discriminating, and a classify- ing eye. He must always bear in mind the wide variety of situations that can lead to war. If he does, he will draw the outline of its salient features in such a way that it can accommodate both the dictates of the age, and those of the immediate situation.

We can thus only say that the aims a belligerent adopts, and the resources he employs, must be governed by the particular character- istics of his own position; but they will also conform to the spirit of the age and to its general character. Finally, they must always be governed by the general conclusions to be drawn from the nature of war itself.

CHAPTER 4

CLOSER DEFINITION OF THE MILITARY OBJECTIVE: THE DEFEAT OF THE ENEMY

THE aim of war should be what its very concept implies—to defeat the enemy. We take that basic proposition as our starting point.

But what exactly does 'defeat' signify? The conquest of the whole of the enemy's territory is not always necessary. If Paris had been taken in 1792 the war against the Revolution would almost certainly for the time being have been brought to an end. There was no need even for the French armies to have been defeated first, for they were not in those days particularly powerful. In 1814, on the other hand, even the capture of Paris would not have ended matters if Bonaparte had still had a sizeable army behind him. But as in fact his army had been largely eliminated, the capture of Paris settled everything in 1814 and again in 1815. Again, if in 1812 Bonaparte had managed, before or after taking Moscow, to smash the Russian army, 120,000 strong, on the Kaluga road, just as he smashed the Austrians in 1805 and the Prussians the following year, the fact that he held the capital would probably have meant that he could make peace in spite of the enormous area still unoccupied. In 1805 Austerlitz* was decisive. The possession of Vienna and two-thirds of the Austrian territory had not sufficed to bring about a peace. On the other hand, after Austerlitz the fact that Hungary was still intact did nothing to prevent peace being made. The final blow required was to defeat the Russian army; the Czar* had no other near at hand and this victory would certainly have led to peace. Had the Russian army been with the Austrians on the Danube in 1805 and shared in their defeat, it would hardly have been necessary to take Vienna; peace could have been imposed at Linz. Equally, a country's total occupation may not be enough. Prussia in 1807 is a case in point. When the blow against the Russian ally in the uncertain victory of Eylau* was not sufficiently decisive, the decisive victory of Friedland* had to be gained in order to achieve what Austerlitz had accomplished the year before.

These events are proof that success is not due simply to general causes. Particular factors can often be decisive—details only known to those who were on the spot. There can also be moral factors which never come to light; while issues can be decided by chances and incidents so minute as to figure in histories simply as anecdotes.

What the theorist has to say here is this: one must keep the dominant characteristics of both belligerents in mind. Out of these characteristics a certain centre of gravity develops, the hub of all power and movement, on which everything depends. That is the point against which all our energies should be directed.

Small things always depend on great ones, unimportant on important, accidentals on essentials. This must guide our approach.

For Alexander, Gustavus Adolphus, Charles XII, and Frederick the Great, the centre of gravity was their army. If the army had been destroyed, they would all have gone down in history as failures. In countries subject to domestic strife, the centre of gravity is generally the capital. In small countries that rely on large ones, it is usually the army of their protector. Among alliances, it lies in the community of interest, and in popular uprisings it is the personalities of the leaders and public opinion. It is against these that our energies should be directed. If the enemy is thrown off balance, he must not be given time to recover. Blow after blow must be aimed in the same direction: the victor, in other words, must strike with all his strength and not just against a fraction of the enemy's. Not by taking things the easy way—using superior strength of filch some province, preferring the security of this minor conquest to great success—but by constantly seeking out the centre of his power, by daring all to win all, will one really defeat the enemy.

Still, no matter what the central feature of the enemy's power may be—the point on which your efforts must converge—the defeat and destruction of his fighting force remains the best way to begin, and in every case will be a very significant feature of the campaign.

Basing our comments on general experience, the acts we consider most important for the defeat of the enemy are the following:

1. Destruction of his army, if it is at all significant
2. Seizure of his capital if it is not only the centre of administration but also that of social, professional, and political activity
3. Delivery of an effective blow against his principal ally if that ally is more powerful than he.

Up till now we have assumed—as is generally permissible—that the enemy is a single power. But having made the point that the defeat of the enemy consists in overcoming the resistance concentrated in his centre of gravity, we must abandon this assumption and examine the case when there is more than one enemy to defeat.

If two or more states combine against another, the result is still politically speaking a *single* war. But this political unity is a matter of degree. The question is then whether each state is pursuing an independent interest and has its own independent means of doing so, or whether the interests and forces of most of the allies are subordinate to those of the leader. The more this is the case, the easier will it be to regard all our opponents as a single entity, hence all the easier to concentrate our principal enterprise into one great blow. If this is at all feasible it will be much the most effective means to victory.

I would, therefore, state it as a principle that if you can vanquish all your enemies by defeating one of them, that defeat must be the main objective in the war. In this one enemy we strike at the centre of gravity of the entire conflict.

There are very few cases where this conception is not applicable— where it would not be realistic to reduce several centres of gravity to one. Where this is not so, there is admittedly no alternative but to act as if there were two wars or even more, each with its own object. This assumes the existence of several independent opponents, and consequently great superiority on their part. When this is the case, to defeat the enemy is out of the question.

We must now address ourselves more closely to the question: when is this objective both feasible and sound?

To begin with, our forces must be adequate:

1. To score a decisive victory over the enemy's
2. To make the effort necessary to pursue our victory to the point where the balance is beyond all possible redress.

Next, we must be certain our political position is so secure that this success will not bring further enemies against us who could force us immediately to abandon our efforts against our first opponent.

France could annihilate Prussia in 1806 even if this brought down Russia on her in full force, since she could defend herself against the Russians on Prussian soil. In 1808 she could do the same in Spain against England; but in respect of Austria she could not. By 1809,

France had to reduce her forces in Spain considerably, and would have had to relinquish Spain altogether if she had not already enjoyed a great moral and material advantage over the Austrians.

These three examples call for careful study. One can win the first decision in a case but lose it on appeal and end by having to pay costs as well.

When the strength and capability of armed forces are being calculated, time is apt to be treated as a factor in total strength on the analogy of dynamics. It is assumed in consequence that half the effort or half the total forces could achieve as much in two years as the whole could do in one. This assumption, which rests, sometimes explicitly, sometimes implicitly, at the basis of military planning, is entirely false.

Like everything else in life, a military operation takes time. No one, obviously, can march from Vilna to Moscow in a week; but here there is no trace of that reciprocal relationship between time and energy that we would find in dynamics.

Both belligerents need time; the question is only which of the two can expect to derive *special advantages* from it in the light of his own situation. If the position on each side is carefully considered, the answer will be obvious: it is the weaker side—but thanks to the laws of psychology rather than those of dynamics. Envy, jealousy, anxiety, and sometimes perhaps even generosity are the natural advocates of the unsuccessful. They will win new friends for him as well as weaken and divide his enemies. Time, then, is less likely to bring favour to the victor than to the vanquished. There is a further point to bear in mind. As we have shown elsewhere, the exploitation of an initial victory requires a major effort. This effort must not only be made but be sustained like the upkeep of a great household. Conquered enemy provinces can, of course, bring additional wealth, but they may not always be enough to meet the additional outlay. If they do not, the strain will gradually increase and in the end resources may be exhausted. Time is thus enough to bring about a change unaided.

Could the money and resources that Bonaparte drew from Russia and Poland in 1812 furnish the men by the hundred thousand whom he needed in Moscow to maintain his position there?

But if the conquered areas are important enough, and if there are places in them vital to the areas still in enemy hands, the rot will spread, like a cancer, by itself; and if only that and nothing else happens, the conqueror may well enjoy the net advantage. Time alone will then complete the work, provided that no help comes from outside, and the area that is still unconquered may well fall without more ado. Thus time can become a factor in the conqueror's strength as well; but only on condition that a counterattack on him is no longer possible, that no reversal is conceivable—when indeed this factor is no longer of value since his main objective has been achieved, the culminating crisis is past, and the enemy, in short, laid low.

That chain of argument was designed to show that no conquest can be carried out too quickly, and that to spread it over a *longer period* than the minimum needed to complete it *makes it not less difficult, but more*. If that assertion is correct, it follows equally that if one's strength in general is great enough to make a certain conquest one must also have the strength to do so in a single operation, not by stages. By 'stages' naturally, we do not mean to exclude the minor halts that are needed for reassembling one's forces or for administrative reasons.

We hope to have made it clear that in our view an offensive war requires above all a quick, irresistible decision. If so, we shall have cut the ground from under the *alternative* idea that a slow, allegedly systematic occupation is safer and wiser than conquest by continuous advance. Nonetheless, even those who have followed us thus far may very likely feel that our views have an air of paradox, of contradicting first impressions and of contradicting views that are as deeply rooted as ancient prejudice and that constantly appear in print. This makes it desirable to examine the alleged objections in some detail.

It is of course easier to reach a nearby object than a more distant one. But if the first does not suit our purpose, a pause, a suspension of activity, will not necessarily make the second half of the journey any easier to complete. A short jump is certainly easier than a long one: but no one wanting to get across a wide ditch would begin by jumping half-way.

If the ideas that underlie the concept of so-called methodical offensive operations are examined, we will usually find the following:

1. Capture the enemy fortresses in your path.
2. Accumulate the stores you need.

3. Fortify important points like *depots*, *bridges*, *positions*, and so forth.
4. Rest your troops in winter quarters and rest-camps.
5. Wait for next year's reinforcements.

If you halt an offensive altogether and stop the forward movement in order to make sure of all the above, you allegedly acquire a new base, and in theory revive your strength as though the whole of your country were immediately to your rear and the army's vigour were renewed with each campaign.

All these are admirable aims, and no doubt they could make offensive war easier; but they cannot make its results more certain. They usually camouflage misgivings on the part of the general or vacillation on the part of the government. We shall now try to roll them up from the left flank.

1. Waiting for reinforcements is just as useful to the other side—if not in our opinion more. Besides, a country can naturally raise almost as many troops in one year as in two, for the net increase in the second year will be very small in relation to the whole.
2. The enemy will rest his troops while we are resting ours.
3. Fortifying towns and positions is no business for the army and therefore no excuse for suspending operations.
4. Given the way in which armies are supplied today they need depots more when they are halted than when on the move. So long as the advance goes properly, enemy supplies will fall into our hands and make up for any shortage in barren districts.
5. Reducing an enemy fortress does not amount to halting the offensive. It is a means of strengthening the advance, and though it causes an apparent interruption it is not the sort of case we have in mind: it does not involve a suspension or a reduction of effort. Only circumstances can decide whether the right procedure is a regular siege, a mere investment, or simply to keep some fortress or other under observation. But we can make the general comment that the answer to this question turns on the answer to another; namely whether it would be too risky to press on and leave no more than an investing force behind. If it is not, and if you still have room to deploy your forces, the right course is to delay a regular siege until all offensive movement is complete. It is important,

therefore, not to give way to the idea of quickly securing everything you have taken, for fear you end by missing something more important.

Such a further advance, admittedly, does seem to place in jeopardy the gains already made.

Our belief then is that any kind of interruption, pause, or suspension of activity is inconsistent with the nature of offensive war. When they are unavoidable, they must be regarded as necessary evils, which make success not more but less certain. Indeed, if we are to keep strictly to the truth, when weakness does compel us to halt, a second run at the objective normally becomes impossible; and if it does turn out to be possible it shows that there was no need for a halt at all. When an objective was beyond one's strength in the first place, it will always remain so.

This seems to us to be generally the case. In drawing attention to it we desire only to dispose of the idea that time, in itself, can work for the attacker. But the political situation can change from year to year, and on that account alone there will often be cases to which this generalization does not apply.

We may perhaps appear to have forgotten our initial thesis and only considered offensive war; but this is not so. Certainly a man who can afford to aim at the enemy's total defeat will rarely have recourse to the defensive, the immediate aim of which is the retention of what one has. But we must insist that defence without an active purpose is self-contradictory both in strategy and in tactics, and in consequence we must repeat that within the limits of his strength a defender must always seek to change over to the attack as soon as he has gained the benefit of the defence. So it follows that among the aims of such an attack, which is to be regarded as the real aim of the defence, however significant or insignificant this may be, the defeat of the enemy could be included. There are situations when the general, even though he had that grand objective well in mind, yet preferred to start on the defensive. That this is no mere abstraction is shown by the campaign of 1812. When Emperor Alexander* took up arms he may not have dreamed he would ever completely destroy his enemy—as in the end he did. But would the idea have been absurd? And would it not have been natural in any case for the Russians to adopt the defensive at the outset of the war?

CHAPTER 5

CLOSER DEFINITION OF THE MILITARY OBJECTIVE—CONTINUED: *LIMITED AIMS**

IN the last chapter we stated the defeat of the enemy, assuming it to be at all possible, to be the true, the essential aim of military activity. We now propose to consider what can be done if circumstances rule that out.

The conditions for defeating an enemy presuppose great physical or moral superiority or else an extremely enterprising spirit, an inclination for serious risks. When neither of these is present, the object of military activity can only be one of two kinds: seizing a small or larger piece of enemy territory, or holding one's own until things take a better turn. This latter is normally the aim of a defensive war.

In considering which is the right course, it is well to remember the phrase used about the latter, *waiting until things take a better turn*, which assumes that there is ground for expecting this to happen. That prospect always underlies a 'waiting' war—that is, a defensive war. The offensive—that is exploiting the advantages of the moment—is advisable whenever the future affords better prospects to the enemy than it does to us. A third possibility, perhaps the most usual, arises when the future seems to promise nothing definite to either side and hence affords no grounds for a decision. Obviously, in that case, the offensive should be taken by the side that possesses the political initiative—that is, the side that has an active purpose, the aim for which it went to war. If any time is lost without good reason, the initiator bears the loss.

The grounds we have just defined for choosing offensive or defensive war have nothing to do with the relative strength of the two sides, although one might suppose that to be the main consideration. But we believe that if it were, the wrong decision would result. No one can say the logic of our simple argument is weak; but does it in practice lead to absurd conclusions? Supposing that a minor state is in conflict with a much more powerful one and expects its position to grow weaker every year. If war is unavoidable, should it not make the most of its opportunities before its position gets still worse? In short, it should attack—but not because attack in itself is advantageous (it will on the contrary increase the disparity of strength) but because the smaller party's interest is either to settle the quarrel before conditions deteriorate or at least to acquire some advantages so as to keep its efforts going. No one could consider this a ludicrous argument. But if the smaller state is quite certain its enemy will attack, it can and should stand on the defensive, so as to win the first advantage. By doing so, it will not be placed at any disadvantage because of the passage of time.

Again, suppose a small power is at war with a greater one, and that the future promises nothing that will influence either side's decisions. If the political initiative lies with the smaller power, it should take the military offensive. Having had the nerve to assume an active role against a stronger adversary, it must do something definite—in other words, attack the enemy unless he obliges it by attacking first. Waiting would be absurd, unless the smaller state had changed its political decision at the moment of executing its policy. That is what often happens, and partly explains why the indeterminate character of some wars leaves a student very much perplexed.

Our discussion of the limited aim suggests that two kinds of limited war are possible: offensive war with a limited aim, and defensive war. We propose to discuss them in separate chapters. But first there is a further point to consider.

The possibility that a military objection can be modified is one we have treated hitherto as deriving only from domestic arguments, and we have considered the nature of the political aim only to the extent that it has or does not have an active content. From the point of view of war itself, no other ingredient of policy is relevant at all. Still, as we argued in the second chapter of Book One (purpose and means in war), the nature of the political aim, the scale of demands put forward by either side, and the total political situation of one's own side, are all factors that in practice must decisively influence the conduct of war. We therefore intend to give them special attention in the following chapter.

CHAPTER 6

A. THE EFFECT OF THE POLITICAL AIM ON THE MILITARY OBJECTIVE

ONE country may support another's cause, but will never take it so seriously as it takes its own. A moderately-sized force will be sent to its help; but if things go wrong the operation is pretty well written off, and one tries to withdraw at the smallest possible cost.

It is traditional in European politics for states to make offensive and defensive pacts for mutual support—though not to the point of fully espousing one another's interests and quarrels. Regardless of the purpose of the war or the scale of the enemy's exertions, they pledge each other in advance to contribute a fixed and usually modest force. A country that makes this sort of alliance does not consider itself thereby involved in actual war with anyone, for that would require a formal declaration and would need a treaty of peace to end it. But even that has never been clearly settled, and practice in the matter varies.

It would all be tidier, less of a theoretical problem, if the contingent promised—ten, twenty, or thirty thousand men—were placed entirely at the ally's disposal and he were free to use it as he wished. It would then in effect be a hired force. But that is far from what really happens. The auxiliary force usually operates under its own commander; he is dependent only on his own government, and the objective the latter sets him will be as ambiguous as its aims.

But even when both states are in earnest about making war upon the third, they do not always say, 'we must treat this country as our common enemy and destroy it, or we shall be destroyed ourselves.' Far from it: the affair is more often like a business deal. In the light of the risks he expects and the dividend he hopes for, each will invest about 30,000 to 40,000 men, and behave as if that were all he stood to lose.

Nor is that attitude peculiar to the case where one state gives another support in a matter of no great moment to itself. Even when both share a major interest, action is clogged with diplomatic reservations, and as a rule the negotiators only pledge a small and limited contingent, so that the rest can be kept in hand for any special ends the shifts of policy may require.

This used to be the universal way in which an alliance operated. Only in recent times did the extreme danger emanating from Bonaparte, or his own unlimited driving power, force people to act in a natural manner. The old way was a half-and-half affair; it was an anomaly, since in essence war and peace admit of no gradations. Nevertheless, the old way was no mere diplomatic archaism that reason could ignore, but a practice deeply rooted in the frailties and shortcomings of the human race.

Finally, some wars are fought without allies; and, political considerations will powerfully affect their conduct as well.

Suppose one merely wants a small concession from the enemy. One will only fight until some modest *quid pro quo* has been acquired, and a moderate effort should suffice for that. The enemy's reasoning will be much the same. But suppose one party or the other finds he has miscalculated, that he is not, as he had thought, slightly stronger than the enemy, but weaker. Money and other resources are usually running short and his moral impulse is not sufficient for a greater effort. In such a case he does the best he can; he hopes that the outlook will improve although he may have no ground for such hopes. Meanwhile, the war drags slowly on, like a faint and starving man.

Thus interaction, the effort to outdo the enemy, the violent and compulsive course of war, all stagnate for lack of real incentive. Neither side makes more than minimal moves, and neither feels itself seriously threatened.

Once this influence of the political objective on war is admitted, as it must be, there is no stopping it; consequently we must also be willing to wage such minimal wars, which consist in *merely threatening the enemy*, with *negotiations held in reserve*.

This poses an obvious problem for any theory of war that aims at being thoroughly scientific. All imperatives inherent in the concept of a war seem to dissolve, and its foundations are threatened. But the natural solution soon emerges. As the modifying principle gains a hold on military operations, or rather, as the incentive fades away, the active element gradually becomes passive. Less and less happens, and guiding principles will not be needed. The art of war will shrivel into prudence, and its main concern will be to make sure the delicate balance is not suddenly upset in the enemy's favour and the half-hearted war does not become a real war after all.

B. WAR IS AN INSTRUMENT OF POLICY

UP to now we have considered the incompatibility between war and every other human interest, individual or social—a difference that derives from human nature, and that therefore no philosophy can resolve. We have examined this incompatibility from various angles so that none of its conflicting elements should be missed. Now we must seek out the unity into which these contradictory elements combine in real life, which they do by partly neutralizing one another. We might have posited that unity to begin with, if it had not been necessary to emphasize the contradictions with all possible clarity and to consider the different elements separately. This unity lies in *the concept that war is only a branch of political activity; that it is in no sense autonomous.*

It is, of course, well-known that the only source of war is politics —the intercourse of governments and peoples; but it is apt to be assumed that war suspends that intercourse and replaces it by a wholly different condition, ruled by no law but its own.

We maintain, on the contrary, that war is simply a continuation of political intercourse, with the addition of other means. We deliberately use the phrase 'with the addition of other means' because we also want to make it clear that war in itself does not suspend political intercourse or change it into something entirely different. In essentials that intercourse continues, irrespective of the means it employs. The main lines along which military events progress, and to which they are restricted, are political lines that continue throughout the war into the subsequent peace. How could it be otherwise? Do political relations between peoples and between their governments stop when diplomatic notes are no longer exchanged? Is war not just another expression of their thoughts, another form of speech or writing? Its grammar, indeed, may be its own, but not its logic.

If that is so, then war cannot be divorced from political life; and whenever this occurs in our thinking about war, the many links that connect the two elements are destroyed and we are left with something pointless and devoid of sense.

This conception would be ineluctable even if war were total war,* the pure elements of enmity unleashed. All the factors that go to make up war and determine its salient features—the strength and allies of each antagonist, the character of the peoples and their governments, and so forth, all the elements listed in the first chapter of Book I—are these not all political, so closely connected with political activity that it is impossible to separate the two? But it is yet more vital to bear all this in mind when studying actual practice. We will then find that war does not advance relentlessly toward the absolute, as theory would demand. Being incomplete and self-contradictory, it cannot follow its own laws, but has to be treated as a part of some other whole; the name of which is policy.

In making use of war, policy evades all rigorous conclusions proceeding from the nature of war, bothers little about ultimate possibilities, and concerns itself only with immediate probabilities. Although this introduces a high degree of uncertainty into the whole business, turning it into a kind of game, each government is confident that it can outdo its opponent in skill and acumen.

So policy converts the overwhelmingly destructive element of war into a mere instrument. It changes the terrible battle-sword that a man needs both hands and his entire strength to wield, and with which he strikes home once and no more, into a light, handy rapier—sometimes just a foil for the exchange of thrusts, feints and parries.

Thus the contradictions in which war involves that naturally timid creature, man, are resolved; if this is the solution we choose to accept.

If war is part of policy, policy will determine its character. As policy becomes more ambitious and vigorous, so will war, and this may reach the point where war attains its absolute form. If we look at war in this light, we do not need to lose sight of this absolute: on the contrary, we must constantly bear it in mind.

Only if war is looked at in this way does its unity reappear; only then can we see that all wars are things of the *same* nature; and this alone will provide the right criteria for conceiving and judging great designs.

Policy, of course, will not extend its influence to operational details. Political considerations do not determine the posting of guards or the employment of patrols. But they are the more influential in the planning of war, of the campaign, and often even of the battle.

That is why we felt no urge to introduce this point of view at the start. At the stage of detailed study it would not have been much help and might have been distracting. But when plans for a war or a campaign are under study, this point of view is indispensable.

Nothing is more important in life than finding the right standpoint for seeing and judging events, and then adhering to it. One point and *one only* yields an integrated view of all phenomena; and only by holding to that point of view can one avoid inconsistency.

If planning a war precludes adopting a dual or multiple point of view—that is, applying first a military, then an administrative eye, then a political, and so on—the question arises whether *policy* is bound to be given precedence over everything.

It can be taken as agreed that the aim of policy is to unify and reconcile all aspects of internal administration as well as of spiritual values, and whatever else the moral philosopher may care to add. Policy, of course, is nothing in itself; it is simply the trustee for all these interests against other states. That it can err, subserve the ambitions, private interests, and vanity of those in power, is neither here nor there. In no sense can the art of war ever be regarded as the preceptor of policy, and here we can only treat policy as representative of all interests of the community.

The only question, therefore, is whether, when war is being planned, the political point of view should give way to the purely military (if a purely military point of view is conceivable at all): that is, should it disappear completely or subordinate itself, or should the political point of view remain dominant and the military be subordinated to it?

That the political view should wholly cease to count on the outbreak of war is hardly conceivable unless pure hatred made all wars a struggle for life and death. In fact, as we have said, they are nothing but expressions of policy itself. Subordinating the political point of view to the military would be absurd, for it is policy that has created war. Policy is the guiding intelligence and war only the instrument, not vice versa. No other possibility exists, then, than to subordinate the military point of view to the political.

If we recall the nature of actual war, if we remember the argument in Chapter 3 above—that *the probable character and general shape of any war should mainly be assessed in the light of political factors and conditions*—and that war should often (indeed today one might say *normally*) be conceived as an organic whole whose parts cannot be separated, so that each individual act contributes to the whole and itself originates in the central concept, then it will be perfectly clear and certain that the supreme standpoint for the conduct of war, the point of view that determines its main lines of action, can only be that of policy.

It is from this point of view, then, that plans are cast, as it were, from a mould. Judgement and understanding are easier and more natural; convictions gain in strength, motives in conviction, and history in sense.

From this point of view again, no conflict need arise any longer between political and military interests—not from the nature of the case at any rate—and should it arise it will show no more than lack of understanding. It might be thought that policy could make demands on war which war could not fulfil; but that hypothesis would challenge the natural and unavoidable assumption that policy knows the instrument it means to use. If policy reads the course of military events correctly, it is wholly and exclusively entitled to decide which events and trends are best for the objectives of the war.

In short, at the highest level the art of war turns into policy—but a policy conducted by fighting battles rather than by sending diplomatic notes.

We can now see that the assertion that a major military development, or the plan for one, should be a matter for *purely military* opinion is unacceptable and can be damaging. Nor indeed is it sensible to summon soldiers, as many governments do when they are planning a war, and ask them for *purely military advice*. But it makes even less sense for theoreticians to assert that all available military resources should be put at the disposal of the commander so that on their basis he can draw up purely military plans for a war or a campaign. It is in any case a matter of common experience that despite the great variety and development of modern war its major lines are still laid down by governments; in other words, if we are to be technical about it, by a purely political and not a military body.

This is as it should be. No major proposal required for war can be worked out in ignorance of political factors; and when people talk, as they often do, about harmful political influence on the management of war, they are not really saying what they mean. Their quarrel should be with the policy itself, not with its influence. If the policy is

right—that is, successful—any intentional effect it has on the conduct of the war can only be to the good. If it has the opposite effect
the policy itself is wrong.

Only if statesmen look to certain military moves and actions to
produce effects that are foreign to their nature do political decisions
influence operations for the worse. In the same way as a man who
has not fully mastered a foreign language sometimes fails to express
himself correctly, so statesmen often issue orders that defeat the
purpose they are meant to serve. Time and again that has happened,
which demonstrates that a certain grasp of military affairs is vital for
those in charge of general policy.

Before continuing, we must guard against a likely misinterpretation. We are far from believing that a minister of war immersed in
his files, an erudite engineer or even an experienced soldier would,
simply on the basis of their particular experience, make the best
director of policy—always assuming that the prince himself is not in
control. Far from it. What is needed in the post is distinguished
intellect and strength of character. He can always get the necessary
military information somehow or other. The military and political
affairs of France were never in worse hands than when the brothers
Belle-Isle* and the Duc de Choiseul* were responsible— good soldiers
though they all were.

If war is to be fully consonant with political objectives, and policy
suited to the means available for war, then unless statesman and
soldier are combined in one person, the only sound expedient is to
make the commander-in-chief a member of the cabinet, so that the
cabinet can share in the major aspects of his activities.* But that, in
turn, is only feasible if the cabinet—that is, the government—is near
the theatre of operations, so that decisions can be taken without
serious loss of time. That is what the Austrian Emperor did in 1809,*
and the allied sovereigns in 1813–1815. The practice justified itself
perfectly.

What is highly dangerous is to let any soldier but the commander-
in-chief exert an influence in cabinet. It very seldom leads to sound
vigorous action. The example of France between 1793 and 1795,
when Carnot ran the war from Paris,* is entirely inapplicable, for
terror can be used as a weapon only by a revolutionary government.

Let us conclude with some historical observations.

In the last decade of the eighteenth century, when that remarkable change in the art of war took place, when the best armies saw part of their doctrine become ineffective and military victories occurred on a scale that up to then had been inconceivable, it seemed that all mistakes had been military mistakes. It became evident that the art of war, long accustomed to a narrow range of possibilities, had been surprised by options that lay beyond this range, but that certainly did not go against the nature of war itself.

Those observers who took the broadest view ascribed the situation to the general influence that policy had for centuries exerted, to its serious detriment, on the art of war, turning it into a half-and-half affair and often into downright make-believe. The facts were indeed as they saw them; but they were wrong to regard them as a chance development that could have been avoided. Others thought the key to everything was in the influence of the policies that Austria, Prussia, England and the rest were currently pursuing.

But is it true that the real shock was military rather than political? To put it in the terms of our argument, was the disaster due to the effect of policy on war, or was policy itself at fault?

Clearly the tremendous effects of the French Revolution abroad were caused not so much by new military methods and concepts as by radical changes in policies and administration, by the new character of government, altered conditions of the French people, and the like. That other governments did not understand these changes, that they wished to oppose new and overwhelming forces with customary means: all these were political errors. Would a purely military view of war have enabled anyone to detect these faults and cure them? It would not. Even if there really had existed a thoughtful strategist capable of deducing the whole range of consequences simply from the nature of the hostile elements, and on the strength of these of prophesying their ultimate effects, it would have been quite impossible to act on his speculations.

Not until statesmen had at last perceived the nature of the forces that had emerged in France, and had grasped that new political conditions now obtained in Europe, could they foresee the broad effect all this would have on war; and only in that way could they appreciate the scale of the means that would have to be employed, and how best to apply them.

In short, we can say that twenty years of revolutionary triumph were mainly due to the mistaken policies of France's enemies.

It is true that these mistakes became apparent only in the course of the wars, which thoroughly disappointed all political expectations that had been placed on them. But the trouble was not that the statesmen had ignored the soldiers' views. The military art on which the politicians relied was part of a world they thought was real—a branch of current statecraft, a familiar tool that had been in use for many years. But *that* form of war naturally shared in the errors of policy, and therefore could provide no corrective. It is true that war itself has undergone significant changes in character and methods, changes that have brought it closer to its absolute form. But these changes did not come about because the French government freed itself, so to speak, from the harness of policy; they were caused by the new political conditions which the French Revolution created both in France and in Europe as a whole, conditions that set in motion new means and new forces, and have thus made possible a degree of energy in war that otherwise would have been inconceivable.

It follows that the transformation of the art of war resulted from the transformation of politics. So far from suggesting that the two could be disassociated from each other, these changes are a strong proof of their indissoluble connection.

Once again: war is an instrument of policy. It must necessarily bear the character of policy and measure by its standards. The conduct of war, in its great outlines, is therefore policy itself, which takes up the sword in place of the pen, but does not on that account cease to think according to its own laws.

CHAPTER 7

THE LIMITED AIM: OFFENSIVE WAR

EVEN when we cannot hope to defeat the enemy totally, a direct and positive aim still is possible: the occupation of part of his territory.

The point of such a conquest is to reduce his national resources. We thus reduce his fighting strength and increase our own. As a result we fight the war partly at his expense. At the peace negotiations, moreover, we will have a concrete asset in hand, which we can either keep or trade for other advantages.

This is a very natural view to take of conquered territory, the only drawback being the necessity of defending that territory once we have occupied it, which might be a source of some anxiety.

In the chapter on the culminating point of victory* we dealt at some length with the way in which an offensive weakens the attacking force, and showed how a situation might develop that could give rise to serious consequences.

Capturing enemy territory will reduce the strength of our forces in varying degrees, which are determined by the location of the occupied territory. If it adjoins our own—either as an enclave within our territory or adjoining it—the more directly it lies on the line of our main advance, the less our strength will suffer. Saxony in the Seven Years War was a natural extension of the Prussian theatre, and its occupation by Frederick the Great made his forces stronger instead of weaker; for Saxony is nearer Silesia than it is to the Mark,* and covers both of them.

Even the conquest of Silesia in 1740 and 1741, once completed, was no strain on Frederick's strength on account of its shape and location and the contour of its frontiers. So long as Saxony was not in Austrian hands, Silesia offered Austria only a narrow frontier, which in any case lay on the route that either side would have to take in advancing.

If, on the other hand, the territory taken is a strip flanked by enemy ground on either side, if its position is not central and its configuration awkward, its occupation will become so plain a burden as to make an enemy victory not just easier but perhaps superfluous. Every time the Austrians invaded Provence from Italy they were forced to give it up without any fighting. In 1744 the French thanked God for allowing them to leave Bohemia without having suffered a defeat. Frederick in 1758 found it impossible to hold his ground in Bohemia and Moravia with the same force that had fought so brilliantly the previous year in Silesia and Saxony. Of armies that had to give up some captured territory just because its conquest had so weakened them, examples are so common that we need not trouble to quote any more of them.

The question whether one should aim at such a conquest, then, turns on whether one can be sure of holding it or, if not, whether a temporary occupation (by way of invasion or diversion) will really be worth the cost of the operation and, especially, whether there is any risk of being strongly counterattacked and thrown off balance. In the chapter on the culminating point, we emphasized how many factors need to be considered in each particular case.

Only one thing remains to be said. An offensive of this type is not always appropriate to make up for losses elsewhere. While we are busy occupying one area, the enemy may be doing the same somewhere else. If our project is not of overwhelming significance, it will not compel the enemy to give up his own conquest. Thorough consideration is therefore necessary in order to decide whether on balance we will gain or lose.

In general one tends to lose more from occupation by the enemy than one gains from conquering his territory, even if the value of both areas should be identical. The reason is that a whole range of resources are denied to us. But since this is also the case with the enemy, it ought not to be a reason for thinking that retention is more important than conquest. Yet this is so. The retention of one's own territory is always a matter of more direct concern, and the damage that our state suffers may be balanced and so to speak neutralized only if retaliation promises sufficient advantage—that is to say the gains are substantially greater.

It follows from all this that a strategic attack with a limited objective is burdened with the defence of other points that the attack itself will not directly cover—far more burdened than it would be if aimed at the heart of the enemy's power. The effect is to limit the scale on which forces can be concentrated, both in time and in space.

If this concentration is to be achieved, at least in terms of time, the offensive must be launched from every practicable point at once. Then, however, the attack loses the other advantage of being able to stay on the defensive here and there and thus make do with a much smaller force. The net result of having such a limited objective is that everything tends to cancel out. We cannot then put all our strength into a single massive blow, aimed in accordance with our major interest. Effort is increasingly dispersed; friction everywhere increases and greater scope is left for chance.

That is how events tend to develop, dragging the commander down, frustrating him more and more. The more conscious he is of his own powers, the greater his self-confidence, the larger the forces he commands, then the more he will seek to break loose from this tendency, in order to give some one point a preponderant importance, even if this should be possible only by running greater risks.

CHAPTER 8

THE LIMITED AIM: DEFENSIVE WAR

THE ultimate aim of a defensive war, as we have seen, can never be an absolute negation. Even the weakest party must possess some way of making the enemy conscious of its presence, some means of threatening him.

No doubt that end could in theory be pursued by wearing the enemy down. He has the positive aim, and any unsuccessful operation, even though it only costs the forces that take part in it, has the same effect as a retreat. But the defender's loss is not incurred in vain: he has held his ground, which is all he meant to do. For the defender then, it might be said, his positive aim is to hold what he has. That might be sound if it were sure that a certain number of attacks would actually wear the enemy down and make him desist. But this is not necessarily so. If we consider the relative exhaustion of forces on both sides, the defender is at a *disadvantage*. The attack may weaken, but only in the sense that a turning point may occur. Once that possibility is gone, the defender weakens more than the attacker, for two reasons. For one thing, he is weaker anyway, and if losses are the same on both sides, it is he who is harder hit. Second, the enemy will usually deprive him of part of his territory and resources. In all this we can find no reason for the attacker to desist. We are left with the conclusion that if the attacker sustains his efforts while his opponent does nothing but ward them off, the latter can do nothing to neutralize the danger that sooner or later an offensive thrust will succeed.

Certainly the exhaustion or, to be accurate, the fatigue of the stronger has often brought about peace. The reason can be found in the half-hearted manner in which wars are usually waged. It cannot be taken in any scientific sense as the ultimate, universal objective of all defence.

Only one hypothesis remains: that the aim of the defence must embody the idea of waiting—which is after all its leading feature. The idea implies, moreover, that the situation can develop, that in itself it may improve, which is to say that if improvement cannot be effected from within—that is, by sheer resistance—it can only come from without; and an improvement from without implies a change in the political situation. Either additional allies come to the defender's help or allies begin to desert his enemy.

Such, then, is the defender's aim if his lack of strength prohibits any serious counterattack. But according to the concept of the defence that we have formulated, this does not always apply. We have argued that the defensive is the more effective form of war, and because of this effectiveness it can also be employed to execute a counteroffensive on whatever scale.

These two categories must be kept distinct from the very start, for each has its effect on the conduct of the defence.

The defender's purpose in the first category is to keep his territory inviolate, and to hold it for as long as possible. That will gain him time, and gaining time is the only way he can achieve his aim. The positive aim, the most he can achieve, the one that will get him what he wants from the peace negotiations, cannot yet be included in his plan of operations. He has to remain strategically passive, and the only success he can win consists in beating off attacks at given points. These small advantages can then be used to strengthen other points, for pressure may be severe at all of them. If he has no chance of doing so, his only profit is the fact that the enemy will not trouble him again for a while.

That sort of defence can include minor offensive operations without their altering its nature or purpose. They should not aim at permanent acquisitions but at the temporary seizure of assets that can be returned at a later date. They can take the form of raids or diversions, perhaps the capture of some fortress or other, but always on condition that sufficient forces can be spared from their defensive role.

The second category exists where the defence has already assumed a positive purpose. It then acquires an active character that comes to the fore in proportion as the scale of feasible counterattack expands. To put it in another way: the more the defensive was deliberately chosen in order to make certain of the first round, the more the defender can take risks in laying traps for the enemy. Of these, the boldest and, if it works, the deadliest, is to retire into the interior. Such an expedient, nonetheless, could hardly be more different from the first type of defensive.

One need only think of the difference between Frederick's situation in the Seven Years War and the situation of Russia in 1812. When war broke out, Frederick's readiness for it gave him some advantage. It meant he could conquer Saxony—such a natural extension of his theatre of war that its occupation put no strain upon his forces, but augmented them. In the campaign of 1757 he sought to continue and develop his strategic offensive, which was not impossible so long as the Russians and the French had not arrived in Silesia, the Mark, and Saxony. But the offensive failed; he was thrown back on the defensive for the rest of the campaign, abandoning Bohemia and having to clear his own base of operations of the enemy. That required the use of the same army to deal first with the French and then with the Austrians. What successes he achieved he owed to the defensive.

By 1758, when his enemies had drawn the noose more tightly round him and his forces were becoming seriously outnumbered, he still planned a limited offensive in Moravia; he aimed at seizing Olmütz before his adversaries were in the field. He did not hope to hold it, still less to make it a base for a further advance, but simply to use it as a sort of outwork, as a *contre-approche* against the Austrians, designed to make them spend the rest of the campaign, and possibly a second year's, in trying to retake it. That effort was a failure too, and Frederick now abandoned any thought of a serious offensive, realizing that it would still further reduce his relative strength. A compact position in the centre of his territories, in Silesia and Saxony, exploitation of interior lines for quickly reinforcing any danger point, small raids as opportunities occurred, quietly waiting meanwhile on events so as to economize his strength for better times—such were the main elements in his plans. Gradually, his operations became more passive. Realizing that even victories cost too much, he tried to manage with less. His one concern was to gain time, and hold on to what he had. Less and less was he willing to give ground and he did not scruple to adopt a thorough-going cordon-system; both Prince Henry's positions in Saxony and those of the King in the mountains of Silesia deserve this description. His letters to the Marquis d'Argens* show how keenly he looked forward to winter quarters and how much he hoped he would be able to take them up without incurring serious losses in the meantime.

To censure Frederick for this, and see in his behaviour evidence of low morale, would in our view be a very superficial judgement.

Devices such as the entrenched camp at Bunzelwitz,* the positions that Prince Henry chose in Saxony and the King in the Silesian mountains, may not seem to us today the sort of measure on which to place one's final hope—tactical cobwebs that a man like Bonaparte would soon have cleared away. But one must remember that times have changed, that war has undergone a total transformation and now draws life from wholly different sources. Positions that have lost all value today could be effective then; and the enemy's general character was a factor as well. Methods which Frederick himself discounted could be the highest degree of wisdom when used against the Austrian and Russian forces under men like Daun and Buturlin.*

This view was justified by success. By quietly waiting on events Frederick achieved his goal and avoided difficulties that would have shattered his forces.

At the start of the 1812 campaign, the strength with which the Russians opposed the French was even less adequate than Frederick's at the outset of the Seven Years War. But the Russians could expect to grow much stronger in the course of the campaign. At heart, all Europe was opposed to Bonaparte; he had stretched his resources to the very limit; in Spain he was fighting a war of attrition; and the vast expanse of Russia meant that an invader's strength could be worn down to the bone in the course of five hundred miles' retreat. Tremendous things were possible; not only was a massive counterstroke a certainty if the French offensive failed (and how could it succeed if the Czar* would not make peace nor his subjects rise against him?) but the counterstroke could bring the French to utter ruin. The highest wisdom could never have devised a better strategy than the one the Russians followed unintentionally.

No one thought so at the time, and such a view would have seemed far-fetched; but that is no reason for refusing to admit today that it was right. If we wish to learn from history, we must realize that what happened once can happen again; and anyone with judgement in these matters will agree that the chain of great events that followed the march on Moscow was no mere succession of accidents. To be sure, had the Russians been able to put up any kind of defence of their frontiers, the star of France would probably have waned, and luck would probably have deserted her; but certainly not on that colossal and decisive scale. It was a vast success; and it cost the

Russians a price in blood and perils that for any other country would have been higher still, and which most could not have paid at all.

A major victory can only be obtained by positive measures aimed at a *decision*, never by simply waiting on events. In short, even in the defence, a major stake alone can bring a major gain.

APPENDIX
COMPLETE CONTENTS LIST OF *ON WAR*

EXPLANATORY NOTES

15 *the recent wars*: the continental wars, known as the French Revolutionary Wars, between the period of the French Revolution (1792–1801) and the Napoleonic Wars (1803–15, although Napoleon, still referred to as Bonaparte, fought in the later campaigns of the French Revolutionary Wars).

put him in a situation that is even more unpleasant: this passage became particularly popular with the Games Theorists of the 1960s.

30 *in the chapter on war plans*: the last book of *On War*, omitted here.

37 *Frederick the Great*: Frederick II Hohenzollern, king of Prussia (1712–86), who used a series of wars, mainly against Austria, but also against France and Russia, to increase the Hohenzollern possessions, notably to include Silesia and large parts of Prussia.

Austria: the Holy Roman Empire until 1806, uniting many territories including Austria, Hungary, and Croatia under the rule of the Habsburg dynasty. The emperor of Austria was also, in personal union, king of Hungary. Napoleon forced Emperor Francis Joseph II to resign the crown of the Holy Roman Empire, which was dissolved; henceforth the Habsburgs were reduced to being emperors of Austria and kings of Hungary.

Seven Years War: this war (1756–63) pitted Prussia, Britain, and Hanover against Austria, France, Russia, Saxony, Sweden, and Spain. Britain fought France for supremacy overseas and the British captured French Canada and ousted the French from India. Prussia, under Frederick the Great, fought Austria for domination of Germany. The war was ended in 1763 by the Treaties of Paris and Hubertusburg, leaving Britain the supreme European naval and colonial power and Prussia in a much stronger position in central Europe.

Charles XII: king of Sweden (1682–1718), Charles fought various wars against Denmark, Poland, and Russia, and was defeated by Russia at the famous battle of Poltava that ended Swedish predominance in the Baltic, replacing it with an ascendant Russia (1709).

ad hominem: about the individual, not about generalities.

40 *what cash payment is in commerce*: one of Clausewitz's particularly incisive comparisons from the world of commerce, see also p. 100.

57 *Puységur*: Jacques-François de Chastenet, marquis de Puységur (1655–1743), French marshal and military author of *Art de la guerre, par principes et par règles, ouvrage de M. Le Maréchal [Jacques Francois] de Puységur* (Paris: Charles-Antoine Jombert, 1748).

Marshal Luxembourg: François-Henri de Montmorency, count of

Boutteville, duke of Luxembourg (1628–95), marshal (i.e. commander-in-chief) of France.

59 *Henry IV*: king of France (1553–1610), who famously reunited France after the wars of religion by converting from Protestantism to Catholicism, but was later assassinated. He was the father of Louis XIII.

Newton: Sir Isaac Newton (1643–1727), English mathematician and physicist, whose hugely influential work embraced mechanics, gravitation, optical experiments, and planetary motion. Author of *Principia mathematica* (1687). See Introduction, p. xiii.

Euler: Leonard Euler (1707–83), Swiss mathematician who elucidated the nature of functions and introduced ideas of convergence and rigorous argument into mathematics, see Introduction, pp. xiii f., xxxii.

74 *Chapter 1 of Book I*: actually Chapter 2.

76 *usually known as 'deployment'*: the original German word here is '*Evolution*', meaning the deployment of troops within battle as opposed to general operational manoeuvres.

80 *'art of war' or 'science of war'*: the words 'art' and 'science' have exchanged meanings in English. Art derives from the Latin *ars*, meaning skill, practical capability, and the equivalent German word used by Clausewitz, *Kunst*, comes from *Können*, to be able to do, a word related to the English verb 'can'. By contrast, *Wissenschaft* (science) derives from *scientia*, wisdom in the abstract. Following a similar logic, at some Scottish universities, physics (as opposed to the practical skill of engineering) is still called 'natural philosophy'. See also pp. 98–101.

81 *properties of its instrument*: the armed forces.

83 *One ingenious mind*: Clausewitz here criticizes the Prussian strategist Adam Heinrich Dietrich von Bülow, whose book *Spirit of the new System of War* was a bestseller among military circles at the time: *Geist des neuern Kriegssystems, hergeleitet aus dem Grundsatze einer Basis der Operationen* (Hamburg, 1799). Clausewitz wrote a scathing review of this book, see *Clausewitz: Verstreute kleine Schriften*, ed. Werner Hahlweg, vols. i–iii (Osnabrück: Biblio Verlag, 1979), 65–88.

another geometrical principle was then exalted: the reference is to Antoine Henri Jomini.

97 *a Newton or an Euler*: see notes to p. 59.

a Condé or a Frederick: Louis II de Bourbon, prince of Condé (1621–86), marshal of France under King Louis XIV, fought in many of his wars; Frederick II, the Great, king of Prussia 1712–86, was a notable military figure who fought the Silesian Wars against Austria (1740–63), and a principal in the Seven Years War (1756–63).

98 *Art of War or Science of War*: see note to p. 80.

100 *condottieri*: the Italian captains of mercenary armies used by most sides during the wars of the Italian Renaissance.

105 *oblique order of battle*: a formation developed by Frederick the Great especially for his cavalry.

106 *Prince Louis . . . Rüchel*: Prince Louis, Friedrich Ludwig Christian, prince of Prussia (1772–1806), nephew of Frederick the Great, fought at the battle of Saalfeld a few days before the double battle of Jena and Auerstadt on 14 October 1806. The Prussian generals Tauentzien (Count Friedrich Bogislaw Emanuel Tauentzien von Wittenberg, 1760–1824), Grawert (Julius August Reinhold von Grawert, 1746–1821), and Rüchel (Ernst von Rüchel, 1754–1823) all fought in the battle of Jena, when the French under Napoleon inflicted a crushing defeat on Prussia. Kapellendorf was the site of one of several skirmishes during the battle. Grawert also took part in the 1812 Prussian campaign against Russia, but after 1813 on the side of Russia against Napoleon. Rüchel, along with Clausewitz himself, became a French prisoner of war after Jena.

Hohenlohe: Friedrich Ludwig, prince of Hohenlohe-Ingelfingen (1746–1818), Prussian general who participated in the wars against Revolutionary France and in Prussia's defeat at the battles of Jena and Auerstedt, after which he was taken prisoner by the French.

111 *Archduke Charles*: Charles Louis John, archduke of Austria (1771–1847), general and commentator on the French Revolutionary and Napoleonic Wars. He was defeated by Napoleon in Bavaria in 1809. See Erzherzog Karl von Österreich, *Grundsätze der höhern Kriegskunst für die Generäle der österreichischen Armee* (Vienna: Kaiserliche und Kaiserlich-Königliche Hof. und Staatsdruckerei, 1806).

Austrians: when speaking about the Austrians, what is usually meant are the armed forces of many ethnic groups united in the Holy Roman Empire, later the Austrian Empire, and the Kingdom of Hungary, including Croats, Bohemians, and other European nationalities.

Moreau and Hoche: Jean-Victor Moreau (1763–1813), a general in the French Revolutionary Wars who achieved victory over Austria at the battle of Hohenlinden, but then turned against Napoleon and joined the anti-Napoleonic coalition; he fell at the battle of Dresden; Lazare Hoche (1768–97), French general and Minister for War during the French Revolution.

French Directory: the French revolutionary government constituted in 1795, overthrown by Napoleon in 1799.

Leoben: the Armistice of Leoben (18 April 1797) ended the War of the First Coalition against Revolutionary France (1792–7).

112 *Campo Formio*: the peace of Campo Formio on 18 October 1799 ended the Second Coalition War against Revolutionary France.

113 *Mantua*: the siege of Mantua in Northern Italy (1796) by French forces under Napoleon, who won the city from the Austrians.

Wurmser: Dagobert Wurmser, count (1724–97), Imperial (i.e. Austrian) field marshal.

114 *Louis XIV*: king of France (1638–1715), who reigned from 1643 to
     1715, also known as Louis the Great and the Sun King, champion
     of royal absolutism, through whose wars and diplomatic successes pre-
     revolutionary France reached its greatest extension and a state of
     predominance in Europe.

115 *Blücher ... Mormant*: Gebhard Leberecht von Blücher (1742–1819),
     prince of Wahlstatt, Prussian field marshal, one of the victors of Waterloo.
     Étoges, Champ-Aubert, Montmirail, Montereau, and Mormant were all
     victories for Napoleon in Northern France in 1814: Étoges on 13 and
     14 February, Champ-Aubert, where Napoleon beat the British, on
     10 February, Montmirail on 11 February, Montereau on 18 February,
     and Mormant on 17 February. Karl, prince of Schwarzenberg (1771–
     1820), was an Imperial field marshal in the wars against Napoleon; he
     surrendered to Napoleon at Ulm in October 1805 but managed to dis-
     engage his forces.

116 *Crown Prince of Württemberg*: Wilhelm of Württemberg (king after 1816)
     (1781–1864), allied with Napoleon until 1813 when he joined the coalition
     against him.

     Wittgenstein: Ludwig Adolf Peter, Count Wittgenstein (1769–1843),
     Russian field marshal who participated in the battles of Austerlitz and
     Friedland, and fought against Napoleon in the 1812 campaign.

     Laon and Arcis: the battles of Laon, 9 March 1814, and of Arcis-sur-Aube
     in northern France, 20–1 March 1814, both defeats for Napoleon by the
     Allies during his retreat.

119 *advanced on Moscow*: Napoleon's Russian campaign of 1812 was witnessed
     of course by Clausewitz himself.

     Czar Alexander: Alexander I (1777–1825) fought against Napoleon in
     1805, was defeated at Austerlitz, and concluded the Peace of Tilsit.
     But Napoleon's invasion of 1812 made him the leader of the wars of
     liberation against Napoleon, culminating in Napoleon's defeat at
     Waterloo in 1815.

     Friedland: battle in eastern Germany (14 June 1807) in which Napoleon
     defeated the Russians leading to the peace of Tilsit between Russia and
     France.

     Emperor Francis: Francis II (1768–1835), emperor of the Holy Roman
     Empire and emperor of Austria as Francis I since 1806.

120 *Austerlitz and Wagram*: the battle of Austerlitz in Moravia on 2 December
     1805, during the War of the Third Coalition, where Napoleon won a
     decisive victory over the Austrians and the Russians. Napoleon also
     defeated the Austrians at the battle of Wagram in Austria on 4–6 July
     1809.

124 *General Scharnhorst*: Gerhard Johann David von Scharnhorst (1755–
     1813), Prussian general; see Introduction, pp. vii f., xii–xvi, xviii, xx f.,
     xxix, xxxi.

125 *Tartars, Cossacks and Croats*: Tartars, or Tatars, were the people constituting the Golden Horde, a nomadic people travelling on horseback who conquered most of Russia in the Middle Ages. Cossacks furnished the Russian Empire, and Croats the Habsburg Empire, with irregular forces or partisans, or light cavalry tasked with reconnaissance, to engage with the adversary's stragglers.

127 *Daun's campaigns*: Leopold Joseph, count of Daun, prince of Thiano (1705–66), Imperial field marshal and adversary of Frederick the Great in the Seven Years War.

Feuquières: Antoine-Manassès de Pas, marquis de Feuquières (1648–1711), French general and author of *Mémoires sur la guerre où l'on a rassemblé les maximes les plus nécessaires dans les operations de l'art militaire* (Amsterdam: Francois Chart., 1731). Clausewitz would have had access to a German translation, published in Leipzig in 1738.

129 *War of the Austrian Succession*: several related European conflicts (1740–8) triggered by the death of the emperor Charles VI and the accession of his daughter Maria Theresa in 1740 to the Austrian throne. As a result of the war, Prussia obtained Silesia from Austria, while Maria Theresa kept her throne.

War of the Spanish Succession: a European war (1701–14) provoked by the death of the Spanish king Charles II without issue which marked the end of Louis XIV's attempts to establish French dominance over Europe. Louis XIV managed to bring his Bourbon dynasty to the throne in Spain under the Peace of Utrecht (1713–14), but Spain and France were prevented from being united under one crown.

condottieri: see note to p. 100.

Hannibal: (246–183 BC), ruler of Carthage and opponent of Rome in the Second Punic War, whose invasion of Italy in 218 BC via the Alps and whose routing of the Romans at Cannae have become some of the greatest points of reference in military history.

130 *Code Napoléon*: Napoleon's great legislative work which he exported to his conquered vassal states.

133 *second chapter of Book Two*: this definition is actually first stated in Chapter 1, Book Two, see p. 74.

135 *Alexander*: Alexander the Great, king of Macedonia (356–323 BC), conqueror of Greece, Asia Minor, Persia, and regions up to India, held up by successive generations as a great military example.

136 *Charles XII*: see note to p. 37.

137 *Lacy*: Franz Moritz, count of Lacy (1725–1801), Imperial field marshal who had fought Frederick the Great in the Seven Years War.

Liegnitz: at the battle of Liegnitz in Silesia on 15 August 1760 during the Seven Years War, Frederick II managed to avoid an encirclement by Austrian and Russian forces under Daun and Laudon along the Katzbach river.

141 *Chapter Three of Book Two*: actually Book 1.

145 *the Vendée*: the counter-revolutionary uprising in this French region, 1793–6, bloodily suppressed by the Revolutionary forces.

Prince Eugène and Marlborough: Eugene François, prince of Savoy-Carignan, margrave of Saluzzo (1663–1736), Imperial general, fought successfully against the Turks; John Churchill, duke of Marlborough (1650–1722), British commander-in-chief and statesman, was the victor at the battle of Blenheim (Höchstädt) in the Wars of the Spanish Succession.

146 *A regular army fighting another regular army*: here Clausewitz contrasts symmetric and asymmetric warfare.

Alexander: see note to p. 135.

Caesar: Gaius Julius (100–44 BC), the famous Roman general and statesman. This is the only reference to him in *On War*; his Gallic Wars are never mentioned as source for military history case studies.

Alexander Farnese: (1547–92), Spanish general and governor in the Netherlands under the Habsburgs, renowned for his cruelty.

Gustavus Adolphus: king of Sweden (1594–1632), great military leader in the Thirty Years War, in which he tried to make Sweden the dominant power in Europe, until he fell at the battle of Lützen.

Charles XII: see note to p. 37.

151 *Economy of Force*: note that Clausewitz does *not* advocate using force sparingly.

the last three chapters: the final chapter is not included in this edition.

152 *Chapter Five of Book Two*: this point is actually discussed in Chapter 1 of Book 1, pp. 21–5 above.

153 *revolutionary wars*: the French Revolutionary Wars, which started in 1792, ran into the Napoleonic Wars as the young Bonaparte assumed commands in both.

154 *Chapter Five of Book Two*: see note to p. 152.

160 *Seven Years War*: see note to p. 37.

165 *The campaign of 1812*: Clausewitz had first-hand experience of Napoleon's 1812 campaign in Russia; see Introduction, pp. xxi–xxiii.

172 *Mollwitz, Hohenfriedberg*: the battle of Mollwitz (10 April 1741) during the First Silesian War, won by Frederick II. In the battle of Hohenfriedberg (4 June 1745), during the Second Silesian War, Frederick successfully deployed his oblique order of battle.

Czaslau, Soor, Rossbach: all battles won by Frederick II: Czaslau, or Chotusitz (17 May 1742) during the First Silesian War; Soor in Bohemia (30 September 1745) during the Second Silesian War; and Rossbach (5 November 1757) in the Seven Years War, in which Frederick inflicted a particularly humiliating defeat upon the French.

Bunzelwitz: in Silesia, the encampment of Frederick the Great from 20 August to 25 September 1760 during the Seven Years War, which finally fell to the Russians.

the commander we have been referring to: Frederick II.

174 *Kolin*: a battle near Prague in Bohemia on 8 June 1757 during the Seven Years War, in which the Austrians defeated Frederick II.

177 *Wellington*: Arthur Wellesley, duke of Wellington (1769–1852), British general who fought against France in the Revolutionary Wars and in the Peninsular War of 1811–12; victor at Waterloo, later statesman.

Torres Vedras: fortifications constructed near Lisbon, where Masséna's advance was halted, 1810–11.

Masséna's army: André Masséna, duke of Rivoli, prince of Essling (1758–1817), French marshal, who won victories at Rivoli, Essling, and Wagram.

Cunctator: Quintus Fabius Cunctator (d. 203 BC), Roman general given the name 'the one who hesitates' for his style in battle.

182 *Portugal*: Portugal was invaded and occupied by France during Napoleon's Peninsular War (1808–14).

183 *Leipzig*: a battle in Saxony (16–19 October 1813) in which Napoleon was defeated. This became elevated to the mythical 'Battle of the Nations', a symbol of the awakening of 'German nationalism', although Germans fought on both sides (the Saxons fought with the French against the hated Prussians).

184 *phenomenon of the nineteenth century*: examples include all the popular uprisings against Napoleon, from the famous *guerrilla* in Spain to the uprisings in Tyrol and Prussia, which Clausewitz called 'the people's war'.

197 *the French in Austria . . . the French in Spain*: all these examples refer to the Napoleonic Wars.

198 *even stronger*: at the end of this section, Marie von Clausewitz, the widow of Clausewitz and editor of his work, inserted the following note: 'The manuscript concludes with the passage: "Development of this subject after Book Three, in the essay on the culminating point of victory." An essay by that title has been found in a folder marked "Various Essays: Materials [for a revision of the manuscript]." It appears to be an expansion of the chapter that is merely outlined here, and is printed at the end of Book Seven. Marie von Clausewitz.'

200 *Marengo, Austerlitz, and Jena*: the battle of Marengo (14 June 1800) in the War of the Second Coalition was a decisive French victory of Napoleon's campaign in Italy. For Austerlitz and Jena, see notes to pp. 119 and 106.

201 *Prague*: the battle of Prague (6 May 1757) in the Seven Years War was indecisive and led to the encounter at Kolin (see note to p. 174).

Hohenlinden: a victory for Napoleon on 26 August 1813.

201 *Katzbach*: or Liegnitz (15 August 1813) in which Frederick II encountered the Austrians under Daun and Laudon.

Chapter Twelve of Book Four: omitted from this edition.

206 *Louis XIV*: see note to p. 114.

209 *The Culminating Point of Victory*: compare Book 7 Chapters 4 and 5 above.

216 *In Silesia ... in Bohemia*: during the Silesian Wars, Daun pressed Frederick harder in combat in the areas belonging to the Holy Roman Empire—notably Bohemia—than in the disputed area of Silesia and the independent kingdom of Saxony.

223 *Absolute War and Real War*: this is where Clausewitz fully develops his contrasting concepts of 'absolute war', i.e. war 'absolved' from all physical constraints such as friction, limitations imposed by lacking resources, and above all, political limitations, and 'real war', which subsequent strategists often paraphrased as 'limited war'. 'Absolute war' is not the same as the twentieth-century concept of 'total war', involving the total mobilization of society, and following Ernst von Ludendorff and the National Socialists, aiming at the complete annihilation of the enemy population, if possible (implemented by the Germans in the genocide of the Jews, Sinti, and Romany). See also Book 3, Chapter 16.

224 *Alexander*: see note to p. 135.

227 *In 1742, 1744, 1757, and 1758*: references to the Silesian Wars: First Silesian War (1740–2), Second Silesian War (1744–5) and the Seven Years War (1756–63).

Kolin: see note to p. 174.

Prague: see note to p. 201.

230 *Newton*: see note to p. 59.

231 *condottiere*: see note to p. 100.

232 *Henry IV*: see note to p. 59.

233 *Louis XI*: (1423–83), king of France 1461–83 during the Hundred Years War with England.

Charles VIII: king of France (1470–98), who invaded Italy in 1494.

Louis XIV: see note to p. 114.

Ferdinand of Aragon: Ferdinand (1452–1516), also known as Ferdinand the Catholic, joint ruler of Spain with Isabella of Castile, under whose rule the last Arab kingdom on Spanish soil (Granada) fell, the Jews were expelled from Spain, and America was discovered by Christopher Columbus.

Charles V: (1500–58), Holy Roman Emperor who established Habsburg rule over the Netherlands and united all the Habsburg territories from Spain to Austria under his rule.

Bohemia: part of the Austrian possessions.

German confederation: created in 1815 after the defeat of Napoleon and the end of French occupation.

234 *Gustavus Adolphus*: see note to p. 146.

Charles XII: see note to p. 37.

237 *Maria Theresa*: (1717–1780). Because of the misogynist continental laws of succession, Maria Theresa's succession to her father Charles VI, enshrined in the Act of Succession by the law of Pragmatic Sanction, was contested, provoking the War of Austrian Succession (1740–8). The Elector Charles Albert of Bavaria had himself elected Holy Roman Emperor Charles VII in 1742. Frederick II of Prussia used this opportunity to conquer the Habsburg possession of Silesia (the First Silesian War). Charles VII died in 1745, and Maria Theresa's husband Francis I Stephen of Lorraine was elected Holy Roman Emperor, while Maria Theresa was technically merely his consort, the Austrian archduchess and queen of Hungary and Bohemia. In reality it was she who was the Habsburg heiress, and who ruled the Holy Roman Empire.

241 *Austerlitz*: see note to p. 120.

the Czar: see note to p. 119.

Eylau: the battle of Eylau (8 February 1807) was a victory for Napoleon.

Friedland: see note to p. 119.

247 *Emperor Alexander*: see note to p. 135.

248 *Limited Aims*: limited political aims in war, according to Clausewitz, imposed limitations on it. Strategists of the twentieth century derived from this the concept of 'limited war'.

252 *total war*: these words, chosen in several places by the translators, do not reflect Clausewitz's thinking: he did not think in terms of twentieth-century genocidal Total War, as defined, for example, by Ludendorff (see note on p. 223 above). The original reads: 'wenn der Krieg ganz Krieg, ganz das ungebundene Element der Feindschaft wäre . . .' and would be better translated as 'if war became entirely war', i.e. were deprived of political and other restraints.

256 *Belle-Isle*: Charles-Louis-Auguste Fouquet, duke of Belle-Isle (1684–1761), French marshal, and then Minister of War under Louis XV, and his brother Louis Charles Armand de Belle-Isle (1693–1746); both had influential roles in France's military history under the Ancien Régime.

Duc de Choiseul: Étienne François, duc de Choiseul, marquis de Stainville (1719–85), French statesman, Minister of Foreign Affairs, and then Minister of War to Louis XV.

major aspects of his activities: the first edition has: 'so bleibt . . . nur ein gutes Mittel übrig, nämlich den obersten Feldherrn zum Mitglied des Kabinets zu machen, damit dasselbe Theil an den Hauptmomenten seines Handelns nehme.' In the second edition, which appeared in 1853, the last part of the sentence was changed to: 'damit er in den wichtigsten

Momenten an dessen Beratungen und Beschlüssen teilnehme.' In his 1943 translation, based on the second or on a still later edition, O. J. M. Jolles rendered this alteration correctly as: 'that he may take part in its councils and decisions on important occasions.' That, of course, is a reversal of Clausewitz's original sense. By writing that the commander-in-chief must become a member of the cabinet so that the cabinet can share in the major aspects of his activities, Clausewitz emphasizes the cabinet's participation in military decisions, not the soldier's participation in political decisions.

Of the several hundred alterations of the text that were introduced in the second edition of *On War*, and became generally accepted, this is probably the most significant change. (Translators.)

256 *That is what the Austrian Emperor did in 1809*: Francis II (1768–1835), Holy Roman Emperor, and emperor of Austria as Francis I since 1804.

when Carnot ran the war from Paris: Lazare Nicolas Marguerite Carnot (1753–1823), mathematician, engineer, and politician, Minister of War under the French Revolution, is generally seen as the father of the French *levée en masse* and the concept of the mobilization of the entire population of France for the war effort—an ideal in fact never achieved.

259 *the chapter on the culminating point of victory*: Book 7, Chapter 5.

the Mark: the marches of Brandenburg (Mark Brandenburg) were the most important territorial possession of the Hohenzollerns, who were margraves of Brandenburg (and as such among the electors of the Holy Roman Emperor, hence elector of Brandenburg) before acquiring remote Prussia and assuming the title of kings of Prussia. Berlin lies in the heart of Brandenburg, Prussia was only ever a remote appendage, but was used later as *pars pro toto* (part for the whole) because being 'kings of Prussia' seemed one step better than mere electors or margraves of Brandenburg.

263 *Marquis d'Argens*: French author and confidant of Frederick, who resided in Prussia during the Seven Years War.

264 *Bunzelwitz*: see note to p. 172.

Daun and Buturlin: Alexander Borissovitch, Count Buturlin (1704–67), Russian general field marshal who fought against Prussia in the Seven Years War. For Daun, see note to p. 127.

the Czar: see note to p. 119.

INDEX

Page numbers in italics refer to the Explanatory Notes.

Bhagavad Gita

The Bible Authorized King James Version
  With Apocrypha

Dhammapada

Dharmasūtras

The Koran

The Pañcatantra

The Sauptikaparvan (from the
  Mahabharata)

The Tale of Sinuhe and Other Ancient
  Egyptian Poems

The Qur'an

Upaniṣads

ANSELM OF CANTERBURY	The Major Works
THOMAS AQUINAS	Selected Philosophical Writings
AUGUSTINE	The Confessions On Christian Teaching
BEDE	The Ecclesiastical History
HEMACANDRA	The Lives of the Jain Elders
KĀLIDĀSA	The Recognition of Śakuntalā
MANJHAN	Madhumalati
ŚĀNTIDEVA	The Bodhicaryàvatàra

Late Victorian Gothic Tales

JANE AUSTEN
Emma
Mansfield Park
Persuasion
Pride and Prejudice
Selected Letters
Sense and Sensibility

MRS BEETON
Book of Household Management

MARY ELIZABETH
BRADDON
Lady Audley's Secret

ANNE BRONTË
The Tenant of Wildfell Hall

CHARLOTTE BRONTË
Jane Eyre
Shirley
Villette

EMILY BRONTË
Wuthering Heights

ROBERT BROWNING
The Major Works

JOHN CLARE
The Major Works

SAMUEL TAYLOR
COLERIDGE
The Major Works

WILKIE COLLINS
The Moonstone
No Name
The Woman in White

CHARLES DARWIN
The Origin of Species

THOMAS DE QUINCEY
The Confessions of an English
    Opium-Eater
On Murder

CHARLES DICKENS
The Adventures of Oliver Twist
Barnaby Rudge
Bleak House
David Copperfield
Great Expectations
Nicholas Nickleby
The Old Curiosity Shop
Our Mutual Friend
The Pickwick Papers

*The
Oxford
World's
Classics
Website*

www.oup.com/uk/worldsclassics

- Information about new titles
- Explore the full range of Oxford World's Classics
- Links to other literary sites and the main OUP webpage
- Imaginative competitions, with bookish prizes
- Articles by editors
- Extracts from Introductions
- Special information for teachers and lecturers

www.oup.com/uk/worldsclassics

American Literature

Authors in Context

British and Irish Literature

Children's Literature

Classics and Ancient Literature

Colonial Literature

Eastern Literature

European Literature

History

Medieval Literature

Oxford English Drama

Poetry

Philosophy

Politics

Religion

The Oxford Shakespeare

A complete list of Oxford World's Classics, including Authors in Context, Oxford English Drama, and the Oxford Shakespeare, is available in the UK from the Marketing Services Department, Oxford University Press, Great Clarendon Street, Oxford OX2 6DP, or visit the website at www.oup.com/uk/worldsclassics.

In the USA, visit www.oup.com/us/owc for a complete title list.

Oxford World's Classics are available from all good bookshops. In case of difficulty, customers in the UK should contact Oxford University Press Bookshop, 116 High Street, Oxford OX1 4BR.